HIGHER

T0187223

PHYSICS

SECOND EDITION

Paul Chambers
Iain Moore
& Mark Ramsay

DYNAMIC LEARNING

HODDER
GIBSON
AN HACHETTE UK COMPANY

The Publishers would like to thank the following for permission to reproduce copyright material:

Photo credits: p.1 (background) and Section 1 running head image © NASA, (inset left) © Gordo – Fotolia.com, (inset centre) © David Lochhead – Fotolia.com; **p.23** (top left) © Andrew Ward – Fotolia.com, (top right) © Mona Makela – Fotolia.com, (bottom left) © Imagestate Media (John Foxx)/Recreation, sports & travel VOL01, (bottom right) © Tim Fletcher; **p.25** (top left) © S.Martin/ESA via Getty Images, (bottom left) © James Steidl – Fotolia.com, (right) © Paul Chambers; **p.26** © Gordo – Fotolia.com; **p.27** © Walter Quirtmair – Fotolia.com; **p.36** © StockTrek/Photodisc/Getty Images/Science, Technology & Medicine 2 54; **p.46** © David Lochhead – Fotolia.com; **p.47** © EDWARD KINSMAN/SCIENCE PHOTO LIBRARY; **p.50** © NASA; **p.55** © LOREN WINTERS/VISUALS UNLIMITED, INC./SCIENCE PHOTO LIBRARY; **p.59** © Digital Vision/Getty Images/Astronomy & Space DV25; **p.78** © NASA, R. Williams and The Hubble Deep Field Team (STScI); **p.80** © NASA/WMAP; **p.83** (background) and Section 2 running head image © Kim Steele/Photodisc/Getty Images, (inset centre) © MAXIMILIEN BRICE, CERN/SCIENCE PHOTO LIBRARY, (inset right) © Ingram Publishing Limited/Technology Communication Gold Vol 2 CD 6; **p.93** © Paul Chambers; **p.96** (left) © CHARLES D. WINTERS/SCIENCE PHOTO LIBRARY, (right) © MARTYN F. CHILLMAID/SCIENCE PHOTO LIBRARY; **p.100** (top left) © FERMILAB/SCIENCE PHOTO LIBRARY, (bottom left) © SSPL/Getty Images, (top right) © JEAN-LOUP CHARMET/SCIENCE PHOTO LIBRARY; **p.101** © Kim Steele/Photodisc/Getty Images/Science, Technology & Medicine 2 54; **p.102** © Science History Images/Alamy Stock Photo; **p.104** (top) © MAXIMILIEN BRICE, CERN/SCIENCE PHOTO LIBRARY, (bottom left) © CERN/SCIENCE PHOTO LIBRARY, (bottom right) © GORONWY TUDOR JONES, UNIVERSITY OF BIRMINGHAM/SCIENCE PHOTO LIBRARY; **p.105** (top) © DAVID PARKER/SCIENCE PHOTO LIBRARY, (bottom) © MAXIMILIEN BRICE, CERN (https://creativecommons.org/licenses/by-sa/3.0/deed.en); **p.123** © Ingram Publishing Limited/Technology Communication Gold Vol 2 CD 6; **p.125** © ANDREW LAMBERT PHOTOGRAPHY/SCIENCE PHOTO LIBRARY; **p.127** (left) © SINCLAIR STAMMERS/SCIENCE PHOTO LIBRARY, (right) © Stock Connection Blue/Alamy; **p.131** © ANDREW LAMBERT PHOTOGRAPHY/SCIENCE PHOTO LIBRARY; **p.133** © EDWARD KINSMAN/SCIENCE PHOTO LIBRARY; **p.135** (left) © NASA/SCIENCE PHOTO LIBRARY, (right) © DR GARY SETTLES/SCIENCE PHOTO LIBRARY; **p.138** © EDWARD KINSMAN/ SCIENCE PHOTO LIBRARY; **p.141** © Eye Ubiquitous/Alamy; **p.143** © Richard Griffin – Fotolia.com; **p.145** © Greg Kuchik/Photodisc/Getty Images/ Science, Technology & Medicine 2 54; **p.147** © ozeroner – Fotolia.com; **p.151** © David J. Green/Alamy; **p.159** (background) and Section 3 running head image © Arthur S. Aubry/Photodisc/Getty Images/Science, Technology & Medicine 2 54, (inset left) © Bryan Fisher – Fotolia.com, (inset centre) © valdezrl – Fotolia.com, (inset right) © magann – Fotolia.com; **p.162** © V Khotenko/Fotolia.com; **p.164** © Iain Moore; **p.172** © TREVOR CLIFFORD PHOTOGRAPHY/SCIENCE PHOTO LIBRARY; **p.173** © TREVOR CLIFFORD PHOTOGRAPHY/SCIENCE PHOTO LIBRARY; **p.190** © Bryan Fisher- Fotolia.com; **p.195** ©Tony Gable and C Squared Studios/Photodisc/Getty Images/ Musical Instruments 34; **p.203** (top) © Dean Mitchell – Fotolia.com, (centre) © Imagestate Media (John Foxx)/Colour Tech V3059, (bottom) © tigger11th – Fotolia.com; **p.204** (left) © Bert Hickman/Wikipedia Commons (http://en.wikipedia.org/wiki/Breakdown_voltage), (right) © valdezrl – Fotolia.com; **p.205** © PETER MENZEL/SCIENCE PHOTO LIBRARY; **p.207** © RGB Ventures/SuperStock/Alamy Stock Photo; **p.209** © ANDREW LAMBERT PHOTOGRAPHY/SCIENCE PHOTO LIBRARY; **p.212** © magann – Fotolia.com; **p.213** (top) © Science & Society Picture Library/Getty Images, (centre) © Stockbyte/Photolibrary Group Ltd/Big Business SD101, (bottom) © Airbus/Sipa/Shutterstock.

Acknowledgements:

Questions marked with an * are extracted from past SQA exam papers © Scottish Qualifications Authority and are reproduced by kind permission.

Every effort has been made to trace all copyright holders, but if any have been inadvertently overlooked the Publishers will be pleased to make the necessary arrangements at the first opportunity.

Although every effort has been made to ensure that website addresses are correct at time of going to press, Hodder Gibson cannot be held responsible for the content of any website mentioned in this book. It is sometimes possible to find a relocated web page by typing in the address of the home page for a website in the URL window of your browser.

Hachette UK's policy is to use papers that are natural, renewable and recyclable products and made from wood grown in well-managed forests and other controlled sources. The logging and manufacturing processes are expected to conform to the environmental regulations of the country of origin.

Whilst every effort has been made to check the instructions of the practical work in this book, it is still the duty and legal obligation of schools to carry out their own risk assessments.

Orders: please contact Bookpoint Ltd, 130 Park Drive, Milton Park, Abingdon, Oxon OX14 4SE. Telephone: (44) 01235 827827. Fax: (44) 01235 400454. Email education@bookpoint.co.uk. Lines are open 9.00–5.00, Monday to Saturday, with a 24-hour message answering service. Visit our website at www.hoddereducation.co.uk. Hodder Gibson can be contacted directly at hoddergibson@hodder.co.uk

© Paul Chambers, Mark Ramsay, Iain Moore 2019

First published in 2013 © Paul Chambers, Mark Ramsay, Iain Moore
This second edition published in 2019 by
Hodder Gibson, an imprint of Hodder Education
An Hachette UK company
211 St Vincent Street
Glasgow G2 5QY

Impression number	5	4	3	2	1
Year	2023	2022	2021	2020	2019

All rights reserved; no part of this publication may be reproduced, stored in a retrieval system, or transmitted, in any other form or by any means, electronic, mechanical, photocopying, recording or otherwise without either the prior written permission of Hodder Education or a licence permitting restricted copying in the United Kingdom issued by the Copyright Licensing Agency Ltd, www.cla.co.uk

Cover photo © Tatiana Zinchenko/Shutterstock.com
Illustrations by Fakenham Prepress Solutions, Integra Software Services Pvt. Ltd., Pondicherry, India and Peter Lubach at Redmoor Design
Typeset in Minion Pro 11pt by Integra Software Services Pvt. Ltd., Pondicherry, India
Printed in Italy

A catalogue record for this title is available from the British Library

ISBN: 978 1 5104 5770 6

Contents

Preface

This book has been written to support students studying Higher Grade Physics. It follows closely the Higher Course Specification introduced in April 2018. The book addresses the course assessment in Appendix 1 which explores the relevance of physics in everyday life. It does not offer guidance on how to describe an experiment as it was felt this would be better covered in general class teaching.

Throughout the book, the explanations used to describe phenomena are both qualitative and quantitative, and this again reflects the nature of the CfE course. It is hoped that while understanding the nature of the physics involved, our students will also become better communicators of physics.

In addition to the core text, this book also offers the following features:

Worked Examples

Examples of the types of questions used in the external examination are given frequently throughout the book, along with explanations and strategies for structuring answers, step by step. This allows students to work logically through problems and gives them a structure to follow which will help to reduce errors and maximise achievement.

Questions and Consolidation Questions

Questions boxes are provided at regular intervals throughout the text and give students the opportunity to practice calculations. *Consolidation Questions* appear at the end of each chapter, and provide further means for students to answer questions and consolidate knowledge. There are answers to all questions at the back of the book.

Activities and Research Tasks

Activities throughout the text provide opportunities for students to demonstrate their practical skills and undertake further research as they progress through the course. These can be used as homework activities or as part of a diagnostic procedure.

For Interest

The book attempts to follow the spirit of CfE by providing examples and references to applications of physics beyond the prescribed content. This allows students to gain an indication of how the subject would progress if they wished to continue their studies in physics.

Assessment – Question paper and Assignment skills

The *Assessment – Question paper and Assignment skills* appendix outlines the assessment requirements of the course and also shows the mark allocation for the various sections, broken into 'knowledge and understanding' and 'skills'. It also provides detailed information and guidance on how to undertake a piece of small-scale research, covering data collection, dependent, independent and control variables and data handling. It then illustrates detailed *Worked Examples* where uncertainties are dealt with and how these uncertainties can be combined to give a final answer and the associated uncertainty.

Updates and syllabus changes: important note to teachers and students from the publisher

This book covers the current course arrangements for Higher Physics (from 2018).

Please remember that syllabus arrangements change from time to time. We make every effort to update our textbooks as soon as possible when this happens, but – especially if you are using an old copy of this book – it is always advisable to check whether there have been any alterations to the arrangements since this book was printed. You can check the latest arrangements at the SQA website (www.sqa.org.uk), and you can also check for any specific updates to this book at www.hoddereducation.co.uk/HigherScience.

1

Our Dynamic Universe

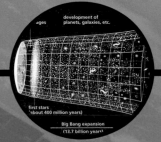

1 Motion

In general terms the word **motion** is used to describe something that is moving. How it is moving can vary dramatically. An object can be moving quickly or slowly. It can be going one way then another like a pendulum or a piston in a car's engine. It can be increasing its velocity or slowing down, flying up in the air or falling, travelling in a circular orbit, a long curvy path or a straight line.

Describing the motion of an object can be complex. This section looks at objects that travel mainly in one dimension and objects that move through the Earth's gravitational field.

In the study of physics, the term 'motion' has a more specific meaning. A physicist would say that motion occurs when an object changes its position. Motion can be described in terms of an object's displacement, velocity, acceleration and time. In this chapter we will consider how objects move in relatively simple situations and analyse the ways in which we describe their motion.

This section falls under the heading of **classical mechanics** which deals with situations where objects are relatively large and relatively slow (in other words, are not travelling at speeds approaching the speed of light). For very small and very rapid objects, classical mechanics does not hold true and **quantum mechanics** is used to describe these situations.

The language of motion

Physicists use specific terms to describe in some detail the motion of objects and this language can take time to become familiar with. We will gradually expand these terms as we work through the text as we will need to be increasingly precise about certain situations so we know exactly the forces that are acting.

> a body – an object
>
> at rest – not moving; motionless
>
> frictionless – having no friction
>
> uniform – constant or not changing

The following sentence shows how such terms can be applied:

'A body of mass 4 kg, at rest, on a frictionless horizontal surface, is acted upon by a force of 20 N.'

It appears an unusual way of saying things but it is more precise than saying a 4 kg object is struck by 20 N.

Describing motion

Displacement is the direct distance from a body's starting position to its final position in an indicated direction. The magnitude (size) of the displacement may be different from the actual distance travelled by the body.

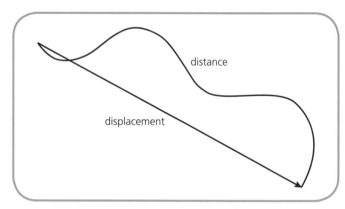

Figure 1.1 Comparing distance and displacement

In the example shown in Figure 1.1, the displacement could be 25 m on a bearing of 120° but the distance travelled would be more like 35 m as the body has not travelled in a straight line.

A person leaving their home, going to work and then coming back to their home will have travelled a large distance but their displacement is zero. They have come back to their starting position.

It may appear unfamiliar to use the term displacement instead of distance but we need to refer to an object's displacement if, for example, it can return to its starting point. An object pushed up a slope could return to its starting point and objects thrown upwards will fall back to their initial position. In such cases it is important to distinguish between distance and displacement.

Worked Example 1.1

1 A body travels 120 m due North, stops then returns along the road a further 30 m. Calculate its displacement.

The magnitude of its displacement is the distance from its starting point to its finishing point. This is 120 m – 30 m. This gives a displacement of 90 m (North). The distance travelled, however, is 150 m.

2 A person walks due North for 270 m then turns right and heads due East for 360 m. Calculate his displacement.

To calculate his displacement we need to consider his direction. We did this in the previous example and this is why we subtracted the 30 m from the 120 m. In this example the direction is at right angles and so we will need to use Pythagoras' theorem or a scale drawing to calculate the displacement.

Pythagoras' theorem states that $x^2 = y^2 + z^2$

$x^2 = 270^2 + 360^2$

$\quad = 72\,900 + 129\,600 = 202\,500$

$x^2 = 202\,500$

$x = 450\,\text{m}$

The distance travelled by the person is 270 m + 360 m = 630 m.
The displacement is 450 m on a bearing of 053°.

Questions

1 Calculate the magnitude of the displacement and the distance travelled in the following situations:

a) An athlete runs two laps of a 440 m track and crosses the line just at the point the laps began.

b) A bus takes the route shown in Figure 1.2.

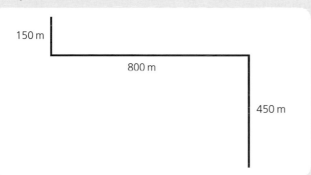

Figure 1.2

c) An object is thrown as shown in Figure 1.3.

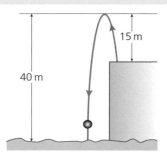

Figure 1.3

d) A geostationary satellite orbits the Earth 10 times at a height of 36 000 km.

e) A ship follows the route as shown in Figure 1.4.

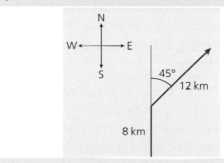

Figure 1.4

f) A trolley rolls down a slope as shown in Figure 1.5 and rebounds to position B.

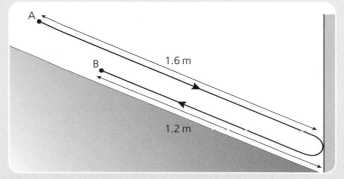

Figure 1.5

Types of motion

Velocity

An object moving from position A to position B has to cover the displacement from A to B in a certain time. The term we use to describe this is its **velocity**. Velocity can be calculated by using the following equation:

$$\text{velocity} = \frac{\text{displacement}}{\text{time}}$$

For example, a cyclist, initially at rest, travels 600 m along a straight road in 120 seconds. Calculate her average velocity.

$$\text{velocity} = \frac{\text{displacement}}{\text{time}}$$
$$= 600/120 = 5\,\text{m s}^{-1}$$

What this means is that for the duration of her journey she travelled 5 m every second (on average).

For most of her journey she would be travelling at a uniform or constant velocity but she would have had to increase her velocity at the beginning of her journey. We use the term average velocity in this context as we do not know her exact velocity at various points of the journey. It is possible to calculate her exact velocity at certain points, such as half way or after 45 seconds, but the calculation is more difficult. It is often referred to as **instantaneous velocity**.

Another way to think of velocity is the 'rate of change' of displacement with respect to time. In other words, the displacement changes by 5 m every second. It may appear slightly contrived but it is merely a different approach to considering the same effect.

This information can also be represented in a graphical form. We can plot a graph of the velocity of the cyclist as the journey progresses. This means we measure the velocity at various stages of the journey: after 5 s, 10 s, 15 s and so on.

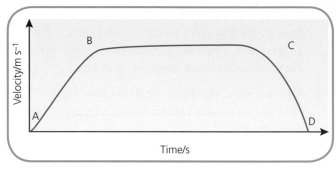

Figure 1.6 A velocity–time graph of a cyclist's journey

Figure 1.6 is known as a velocity–time graph and it shows the velocity of the cyclist during the journey. The graph gives us more information regarding the journey. We can determine the velocity at 20 s or 30 s, for example, or look at the graph to determine when the cyclist was at her fastest, when she was slowing down or when she was travelling at a constant velocity.

By looking at Figure 1.6 we can determine the following:

From A to B the cyclist starts from rest until she reaches her maximum velocity.

From B to C she stays at pretty much the same velocity (uniform velocity).

From C to D she slows down and stops.

Questions

2 A velocity–time graph of the motion of two objects is shown in Figure 1.7.

a) State the maximum velocity in both graphs.

b) At what time(s) was object A's velocity 6 m s^{-1}?

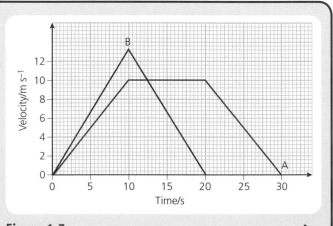

Figure 1.7

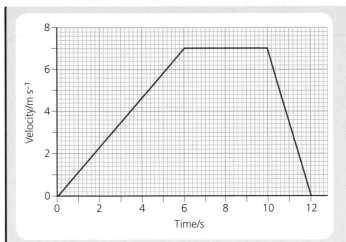

Figure 1.8

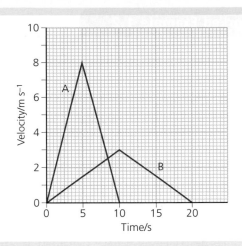

Figure 1.9

3 The motion of an object is represented by the velocity–time graph shown in Figure 1.8.

   a) State the maximum velocity of the object.

   b) What was its velocity at 4 s?

   c) At what time(s) was it travelling at $4\,\mathrm{m\,s^{-1}}$?

4 Velocity–time graphs of two objects are shown in Figure 1.9.

   a) Which object had the greater acceleration?

   b) Explain your answer.

5 Sketch the velocity–time graphs corresponding to the following situations:

   a) A body starts from rest and accelerates uniformly to $5\,\mathrm{m\,s^{-1}}$ in 10 seconds. It remains at that velocity for 4 seconds then it slows down and stops in a further 8 seconds.

   b) A body travelling at $3\,\mathrm{m\,s^{-1}}$ remains at a constant velocity for 5 seconds. It accelerates to $10\,\mathrm{m\,s^{-1}}$ in a further 7 seconds then slows down and stops 8 seconds after that.

   c) A car travelling at $6\,\mathrm{m\,s^{-1}}$ remains at that velocity for 10 seconds then stops in a further 5 seconds. It remains at rest for 5 seconds then accelerates to $8\,\mathrm{m\,s^{-1}}$ in 6 seconds.

Vectors and Scalars

Earlier work in physics will have introduced the difference between vectors and scalars. At Higher the difference has greater significance as we consider objects moving in different directions.

Displacement can be considered the vector equivalent of distance.

Velocity can be considered the vector equivalent of speed.

Acceleration

In describing motion earlier we spoke of objects with constant or changing velocity. **Acceleration** is the term we use when describing an object which has a changing velocity. In everyday language, acceleration refers to an increase in velocity and deceleration to an object with a decreasing velocity. In physics, however, where acceleration refers to changing velocity, it can mean an object increasing or decreasing its velocity. In this textbook, acceleration is the only term we will use.

Acceleration is calculated using the equation:

acceleration = change in velocity/time taken= (final velocity – initial velocity)/time taken

This can be written with symbols as:

$$a = \frac{v - u}{t}$$

Worked Example 1.2

A car starts from rest and reaches a velocity of $30\,\text{m}\,\text{s}^{-1}$ in 10 seconds.

Calculate its acceleration.

$$\text{acceleration} = \frac{\text{change in velocity}}{\text{time taken}}$$

$$= \frac{30 - 0}{10} = 3\,\text{m}\,\text{s}^{-2}$$

The car's velocity changed by $30\,\text{m}\,\text{s}^{-1}$ in 10 seconds. In words, this means that its velocity changed by $3\,\text{m}\,\text{s}^{-1}$ each second. We write this as $3\,\text{m}\,\text{s}^{-1}$ per second. This is sometimes described as '$3\,\text{m}\,\text{s}^{-1}$ (squared)' or written as $3\,\text{m}\,\text{s}^{-2}$. The acceleration of the car is in the same direction as the car's motion. Acceleration is a vector and requires an indication of direction also.

We could describe these data in the form of a velocity–time graph.

Figure 1.10 shows how the velocity of the car changes with time. At this level, we will only consider examples where the acceleration is uniform. There are many examples where the acceleration changes but it can be complex mathematically.

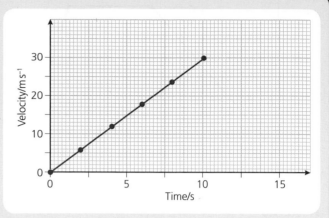

Figure 1.10 A velocity–time graph showing uniform acceleration

In much the same way that we could describe velocity as the rate of change of displacement with respect to time, we can also describe acceleration as the rate of change of velocity with respect to time.

This may seem complex but the principle of the 'rate of change' is an important one in physics and we will return to it throughout the text.

Equations of motion

From earlier work you should be familiar with simple acceleration situations such as a block sliding down a ramp, an object falling under gravity or a car speeding up. As detailed in the previous section, acceleration is the term we use to describe an object when it is increasing or decreasing its velocity. Acceleration is a measure of how quickly an object's velocity changes. We generally calculate acceleration as the change in an object's velocity in one second.

Worked Example 1.3

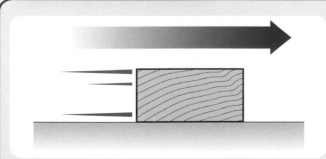

Figure 1.11

A block goes from $4\,\text{m}\,\text{s}^{-1}$ to $10\,\text{m}\,\text{s}^{-1}$ in 3 seconds.

Its velocity goes from $4\,\text{m}\,\text{s}^{-1}$ to $10\,\text{m}\,\text{s}^{-1}$. It increases by $6\,\text{m}\,\text{s}^{-1}$.

It increases by $6\,\text{m}\,\text{s}^{-1}$ in 3 seconds.

This is the same as saying an increase of $2\,\text{m}\,\text{s}^{-1}$ in 1 second. We would describe this as an acceleration of $2\,\text{m}\,\text{s}^{-1}$ per second or $2\,\text{m}\,\text{s}^{-2}$ (2 metres per second squared) in the direction of motion.

→

The equation $a = \frac{v - u}{t}$ can be used in this example if we substitute as follows:

v = final velocity

u = initial velocity

t = time taken for velocity to change

a = acceleration

This gives:

$$a = \frac{v - u}{t} = (10 - 4)/3 = 6/3 = 2\,\mathrm{m\,s^{-2}}$$

Using the equation gives us a quick method of calculating the acceleration. We completed the calculation in one line compared to three or four previously.

The term acceleration is used to describe many everyday situations:

'The sprinter accelerated out of the blocks.'
'The acceleration in this car pulls your head back.'
'The ball accelerated from the moment it left his boot.'

Some descriptions are good but others are misleading and may cause misconceptions when looked at closely.

Further equations

By analysing some simple graphs and rearranging what we have, we can generate additional equations which can help us solve more complicated problems regarding the motion of objects.

In the following equations, these letters will be substituted for the relevant terms:

v = final velocity
u = initial velocity
t = time taken for velocity to change
a = acceleration
s = displacement

Equation 1

$a = \frac{v - u}{t}$ can be manipulated algebraically to give

$$at = v - u$$

Taking u to the other side gives

$$u + at = v$$

Turning around gives

$$v = u + at$$

This is our first equation of motion.

This allows us to calculate any of these terms, a or v for example, if we have information about the others.

Equation 2

The derivation of the second equation comes from the velocity–time graph of a moving object, as shown in Figure 1.12.

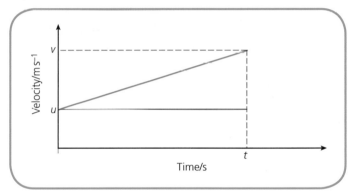

Figure 1.12 The velocity–time graph of an object at velocity u increasing to velocity v after t seconds

As indicated earlier, the magnitude of the displacement is the straight line distance from the starting position to the finishing position. More fully, displacement is the term we give to the distance covered in a certain direction.

The displacement is given by calculating the area of the velocity–time graph for that period. The area can be obtained by splitting the graph into a rectangle at the base and a triangle at the top.

Area of rectangle = length × height = $t \times u$

This can be rearranged to give $u \times t = ut$

Area of triangle = $\frac{1}{2} \times$ length × height

$$= \frac{1}{2} \times t \times (v - u)$$

From equation 1, however, $at = v - u$

Area of triangle $= \frac{1}{2} \times t \times at$

$$= \frac{1}{2} \times at^2$$

$$= \frac{1}{2} at^2$$

The total area (the displacement) can now be obtained by adding the two areas together, therefore

$$s = ut + \frac{1}{2}at^2$$

This is our second equation of motion.

This equation is more complicated than the first as it has four variables, the same term is in two portions of the equation and there is a t squared term to deal with. With simple algebraic manipulation, however, it can be solved in much the same way as equation 1.

Equation 3

This third equation helps us solve problems where we would have to use a combination of equations. It can be quite difficult to follow. Being able to use the equation is the key skill but if you can follow its derivation, it will help with the general algebraic skills required for the course.

Starting with $\qquad v = u + at$

Square both sides $v^2 = (u + at)^2$

Open brackets $\qquad v^2 = u^2 + 2uat + a^2 t^2$

We now rearrange the right-hand side

$$v^2 = u^2 + a(2ut + at^2)$$

Taking the 2 out of the right-hand side bracket

$$v^2 = u^2 + 2a(ut + \frac{1}{2}at^2)$$

As our second equation of motion is

$$s = ut + \frac{1}{2}at^2$$

We can substitute this to get

$$v^2 = u^2 + 2as$$

This is our third equation of motion.

These three equations of motion can be used to solve most of the problems posed in this section. How we combine and use them is the skill that is crucial. In examinations, the equations are given in the data sheet but it saves time if you can remember them as you will spend less time referring to a separate sheet when solving problems.

$$v = u + at \qquad s = ut + \frac{1}{2}at^2 \qquad v^2 = u^2 + 2as$$

Between them these three equations contain five variables: v, u, a, t and s. Combining the equations above with what we know from the information given in the questions can allow us to solve a range of problems.

Worked Example 1.4

1 A car starts from a stationary point and reaches $28\,\mathrm{m\,s^{-1}}$ in 15 seconds.

Calculate its acceleration.

One method of solving such questions is to write the five variables down one side of the page and insert the values that you have been given. This will help you decide which equation to use to calculate the unknown quantity.

$v = 28\,\mathrm{m\,s^{-1}}$
$u = 0\,\mathrm{m\,s^{-1}}$
$a = ?$
$t = 15\,\mathrm{s}$
$s = ?$

To calculate a (acceleration) if we know v, u and t, we should use $v = u + at$.
Inserting the information from the question gives
$28 = 0 + a \times 15$
Rearranging gives
$15a = 28$

$a = 28/15 = 1.87 = 2.0\,\mathrm{m\,s^{-1}}$

You can rearrange the equation prior to inserting the numbers. Some people prefer this and both methods are acceptable.

This would give
$v = u + at$
Rearranging gives
$a = \dfrac{v - u}{t}$
$a = (28 - 0)/15 = 1.87 = 2.0\,\mathrm{m\,s^{-1}}$

2 A ball dropped from a balcony hits the ground 4.0 seconds later.
Calculate its velocity just as it hits the ground and the height it was dropped from.

$v = ?$
$u = 0\,\mathrm{m\,s^{-1}}$
$a = 9.8\,\mathrm{m\,s^{-2}}$ (acceleration due to gravity)
$t = 4.0\,\mathrm{s}$
$s = ?$

→

To calculate v we can use $v = u + at$

This becomes $v = 0 + 9.8 \times 4.0 = 39.2\,\text{ms}^{-1} = 39\,\text{ms}^{-1}$

To calculate s we can use $s = ut + \frac{1}{2}at^2$

This reduces to $s = \frac{1}{2}at^2$ as $u = 0\,\text{ms}^{-1}$.

$s = \frac{1}{2}at^2 = \frac{1}{2} \times 9.8 \times 4.0^2 = 78.4 = 78\,\text{m}$

3 An object moving at $4\,\text{ms}^{-1}$ accelerates to $12\,\text{ms}^{-1}$. It travels a distance of $32\,\text{m}$ during this phase. Calculate its acceleration during this time and how long it takes to reach $12\,\text{ms}^{-1}$.

$v = 12\,\text{ms}^{-1}$
$u = 4\,\text{ms}^{-1}$
$a = ?$
$t = ?$
$s = 32\,\text{m}$

We can use $v^2 = u^2 + 2as$ as we know v, u and s. Substituting gives

$12^2 = 4^2 + 2 \times a \times 32$

$12^2 - 4^2 = 64 \times a$

$144 - 16 = 64 \times a$

$a = 128/64 = 2\,\text{ms}^{-2}$

To calculate t we use $v = u + at$ and this gives

$12 = 4 + 2 \times t$
$t = 4\,\text{s}$

4 An object is dropped from a $22\,\text{m}$ high window. How long does it take to hit the ground and what velocity does it reach just before it hits the ground?

$v = ?$
$u = 0$
$a = 9.8\,\text{ms}^{-2}$
$t = ?$
$s = 22$

To calculate t we use $s = ut + \frac{1}{2}at^2$

$u = 0$ so $ut = 0$

$s = \frac{1}{2}at^2$

$22 = \frac{1}{2} \times 9.8t^2$

$22 = 4.9t^2$

$\sqrt{\frac{22}{4.9}} = t = 2.1\,\text{s}$

To calculate v

$v = u + at$
$= 0 + 9.8 \times 2.1$
$= 20.6 = 21\,\text{ms}^{-1}$

The need for using vectors is more evident in these examples. Objects moving up and down or from side to side involve vector calculations and this is essential for further study.

Questions

6 An object is dropped from an opening $44\,\text{m}$ above the ground.

  a) How long does it take to hit the ground?

  b) What velocity does it reach just before it hits the ground?

7 A ball is dropped from a tower and hits the ground 5 seconds later.

  a) From what height was it dropped?

  b) What is its velocity just before impact?

8 A high-speed train can reach a velocity of $250\,\text{km/hr}$ ($70\,\text{ms}^{-1}$). It takes 3 minutes to reach this velocity from rest.

  a) Calculate its acceleration in ms^{-2}.

  b) What is the minimum distance required for the train to reach this velocity?

9 A sports car can accelerate from rest to 60 mph in 3.2 seconds. Calculate its acceleration in ms^{-2}.

10 A trolley is released from rest on a slope. After 1.3 seconds the trolley passes through a light gate $2.0\,\text{m}$ away.

  a) Calculate its acceleration.

  b) Calculate the velocity of the trolley as it passes the light gate.

11 An object accelerates from rest at $4\,\text{ms}^{-2}$ for 4 seconds.

  a) Calculate its velocity after 4 seconds.

  b) What distance has it travelled after 4 seconds?

  c) How long would it have to maintain this acceleration if it had to cover $100\,\text{m}$?

The term 'acceleration' is used to describe objects which are changing their velocity. Velocity is a **vector** and this requires a direction. Any object which changes direction is changing its velocity and therefore accelerating. This includes cars turning corners and satellites in orbit. You should be familiar with acceleration for objects in a straight line which involves speeding up and slowing down and simple examples such as that of a ball falling.

We notice and 'feel' the acceleration when we are in vehicles that change their velocities rapidly. Fairground rides, motorbikes and rapid stops in cars all make us move unexpectedly and this movement as a result of the acceleration is sometimes not obvious to explain.

Examples to illustrate and further explain acceleration

The following examples give some everyday uses of the term 'acceleration' along with some higher level explanations of the phenomena described.

'The acceleration in this car pulls your head back'

Does the acceleration in a car pull your head back?
If yes, what does the pulling?
If no, why does it go back?

Explanations and answers

The acceleration of a car does not pull your head back. Your head is not attached to anything in the car that can do that, so it is not pulled back. Acceleration is just a description of the motion of a body. It cannot do anything in itself. Saying the acceleration 'pulls your head back' also gives it some sort of life-like quality. This is poor physics and leads to misconceptions.

When the car accelerates it goes forward and increases its velocity. A driver in that car will go forward because of his or her contact with the seat. The driver's head is generally above the level of the seat and not in direct contact with the headrest. The driver's body is accelerated forwards due to the forces applied by the seat. The head has no seat pushing it so it does not move, therefore the body moves forward leaving the head where it was.

'The ball accelerated from the moment it left his boot'

Can the ball accelerate after it has been kicked?

Explanation and answer

It can. Once the ball has left the boot, the external forces on the ball are air resistance and gravity. These forces will make the ball accelerate (change its speed) but it will slow down not speed up.

To our eyes the ball is moving very quickly and, depending on the angle at which it was kicked, it will appear to travel in a straight line in an upwards direction (possibly). It will, however, travel in an arc with the ball slowing down. Technically it is accelerating but not in the way the commentator intended.

'The sprinter accelerated out of the blocks'

This is a reasonably accurate description of what happens. The sprinter is stationary to begin with (at rest) and increases his/her velocity until they reach their maximum running velocity.

Graphs

Graphs are a good way of presenting a lot of data in one image. They can give an overview of the relationship between two variables and make trends or patterns easier to identify. They can also be used to hypothesise what would happen if one variable was to decrease to, for example, zero or increase to a certain value.

In the study of motion we use a number of graphs to analyse the displacement or velocity or acceleration of an object. We plot these variables against time and this allows us to note how each of the properties change as an object's motion is charted.

Velocity–time graphs

The following graphs illustrate how the velocity of an object varies with respect to time.

Worked Example 1.5

A body at rest accelerates uniformly until it reaches a velocity of $12\,\mathrm{m\,s^{-1}}$ in 10 seconds. It then remains at that velocity for a further 5 seconds.

Draw the velocity–time graph for this.

The graph shows how the velocity of the object changes with time. We can also find the velocity at various points along the journey. For example, the velocity after 5 seconds can be obtained by drawing a line vertically from 5 seconds and noting where it crosses the y-axis ($6\,\mathrm{m\,s^{-1}}$).

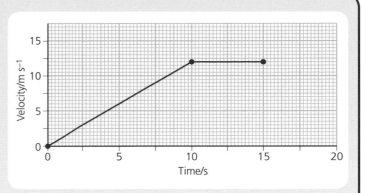

Figure 1.13

Using velocity–time graphs to calculate displacement

Example 1

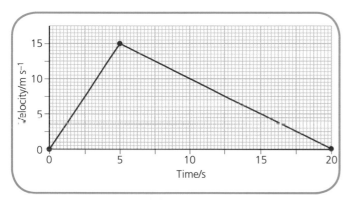

Figure 1.14 The velocity–time graph for an object increasing from rest to $15\,\mathrm{m\,s^{-1}}$ in 5 seconds then slowing down to zero in a further 15 seconds

The displacement can be calculated by working out the area of the velocity–time graph. A simple way to do this is to split this graph into two triangles and calculate the area of each.

Triangle 1: Area $= \frac{1}{2} \times$ base $\times$ height $= \frac{1}{2} \times 5 \times 15$ $= 37.5\,\mathrm{m}$

Triangle 2: Area $= \frac{1}{2} \times$ base $\times$ height $= \frac{1}{2} \times 15 \times 15$ $= 112.5\,\mathrm{m}$

Total displacement $= 150\,\mathrm{m}$

This was a straightforward example but more complex situations can also be examined.

Example 2

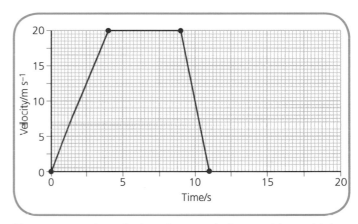

Figure 1.15 The velocity–time graph for an object which starts at rest, then increases its velocity to $20\,\mathrm{m\,s^{-1}}$ in 4 seconds, stays at $20\,\mathrm{m\,s^{-1}}$ for 5 seconds and then stops in a further 2 seconds

The displacement can be calculated by splitting the graph into three simple sections and calculating the areas of the first triangle, the rectangle and the second triangle.

Triangle 1: Area $= \frac{1}{2} \times 4 \times 20 = 40\,\mathrm{m}$

Rectangle: Area $= 20 \times 5 = 100\,\mathrm{m}$

Triangle 2: Area $= \frac{1}{2} \times 2 \times 20 = 20\,\mathrm{m}$

Total displacement $= 40 + 100 + 20 = 160\,\mathrm{m}$

Example 3

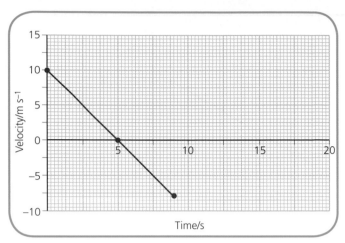

Figure 1.16 The velocity–time graph for an object at $10\,\mathrm{m\,s^{-1}}$ which slows down to $0\,\mathrm{m\,s^{-1}}$ in 5 seconds and then reaches $-8\,\mathrm{m\,s^{-1}}$ in a further 4 seconds

This is more complex as it is a velocity–time graph with a negative velocity. The negative term means that the object is travelling in the opposite direction to its initial movement. An object going away, slowing down, stopping, then returning to you would be a possible description of its motion.

The displacement can be calculated by splitting the graph into two triangles – one above the horizontal line and one below.

Triangle 1: Area $= \frac{1}{2} \times 5 \times 10 = 25\,\mathrm{m}$

Triangle 2: Area $= \frac{1}{2} \times 4 \times -8 = -16\,\mathrm{m}$

The total displacement is $25\,\mathrm{m} - 16\,\mathrm{m} = 9\,\mathrm{m}$. In other words, the object finishes $9\,\mathrm{m}$ away from its initial starting point in the initial direction.

This technique is applicable to all velocity–time graphs. The graphs can be fairly complex in their shape but the principle is the same.

Questions

Velocity–time graphs for a range of moving objects are shown.

12

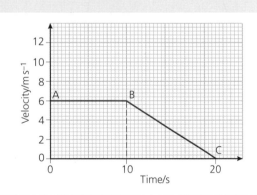

Figure 1.17

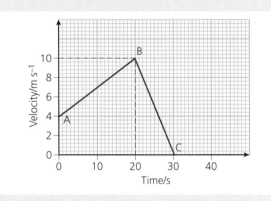

Figure 1.18

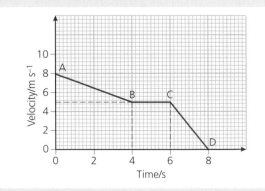

Figure 1.19

a) Calculate the final displacement of the object in Figure 1.17.

b) Calculate its acceleration during section B–C.

13 a) Calculate the acceleration of the object in Figure 1.18 during sections A–B and B–C.

b) Calculate its final displacement.

14 a) How far did the object in Figure 1.19 travel when it was travelling at a constant velocity?

b) Calculate its overall displacement.

c) Calculate the average velocity of the object.

d) Calculate the acceleration during each of the three phases: A–B, B–C and C–D.

15 A trolley is pushed up a slope and a velocity–time graph of its motion is shown in Figure 1.20.

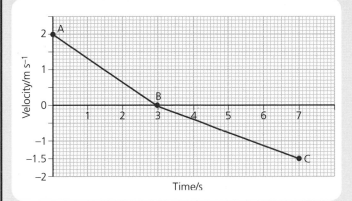

Figure 1.20

a) Calculate the acceleration of the trolley as it travels up the slope.

b) Calculate its acceleration as it moves down the slope.

c) How far up the slope did the trolley travel?

d) Did it return to its starting position? Justify your answer.

e) Calculate its average velocity for the journey.

16 The velocity–time graph for a sky diver is shown in Figure 1.21.

Describe the sky diver's motion and explain the shape of the graph.

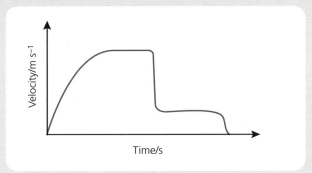

Figure 1.21

17 Many roads now have 'average speed cameras' as a way of reducing speeding in certain areas. They operate by identifying a car at a certain point on a road and then checking that car at a much later point. Are average speed cameras a better way of reducing speeding than traditional speed traps or cameras? Justify your answer.

Displacement–time graphs

These types of graph chart the displacement of an object with time. An object moving steadily from an initial point will have a displacement graph similar to that shown in Figure 1.22. The graph is a simple straight line and it shows that the displacement from its starting point increases by the same amount in the same time.

This is straightforward to interpret and is the displacement–time graph of an object moving steadily.

For an object gradually increasing its velocity, it is best to consider the velocity–time graph and calculate its displacement at certain times along its journey.

Its displacement at certain times can be determined by calculating the area of the graph at these points. These areas will all be triangles and we calculate the area of the triangle from the origin each time.

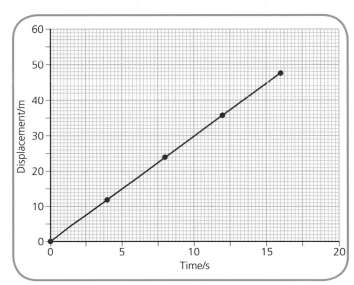

Figure 1.22 A displacement–time graph for an object moving steadily

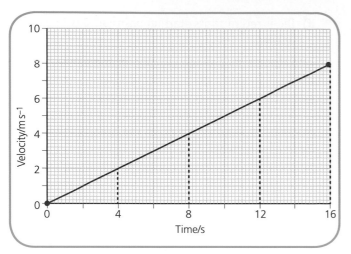

Figure 1.23 The velocity–time graph for an object gradually increasing its velocity

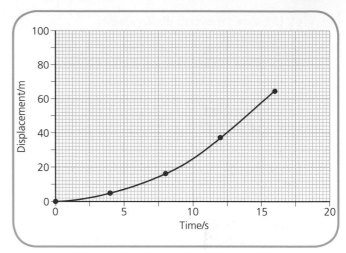

Figure 1.24 A displacement–time graph for the journey shown in Figure 1.23

Area 1 $= \frac{1}{2} \times 4 \times 2 = 4$

Area 2 $= \frac{1}{2} \times 8 \times 4 = 16$

Area 3 $= \frac{1}{2} \times 12 \times 6 = 36$

Area 4 $= \frac{1}{2} \times 16 \times 8 = 64$

We then plot a graph of the displacement against time as shown in Figure 1.24.

It can be seen that the graph is a parabolic shape and similar to the shape of $y = x^2$. This is what we would anticipate as this is a graph of displacement against time.

The equation for displacement is $s = ut + \frac{1}{2}at^2$

and as it starts from rest this becomes $s = \frac{1}{2}at^2$.

This is of a similar form to $y = x^2$, hence the parabolic shape of the graph.

Worked Example 1.6

A car at rest accelerates at $2\,\text{m s}^{-2}$ for 10 seconds.

Draw
a) the velocity–time graph
b) the displacement–time graph.

There are a number of ways to work on this question. One straightforward way is to calculate the velocity after 2 s, 4 s, 6 s, 8 s and 10 s.

Using $v = u + at$ we get $v = 4\,\text{m s}^{-1}$, $8\,\text{m s}^{-1}$, $12\,\text{m s}^{-1}$, $16\,\text{m s}^{-1}$ and $20\,\text{m s}^{-1}$.

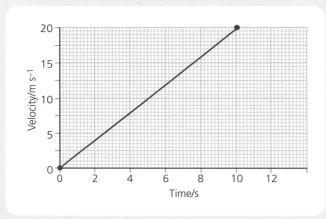

Figure 1.25 The velocity–time graph

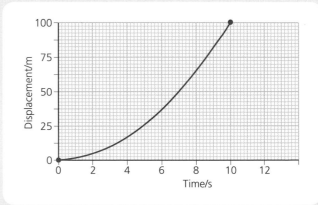

Figure 1.26 The displacement–time graph

This gives a graph as shown in Figure 1.25.

To calculate the displacement we work out the area at 2 s, 4 s, 6 s, 8 s and 10 s. This gives 4 m, 16 m, 36 m, 64 m and 100 m, respectively.

We can also calculate the displacement using
$s = ut + \frac{1}{2}at^2$.

As the car starts from rest we can remove the ut term to give us $s = \frac{1}{2}at^2$.

This gives the displacement as 4 m, 16 m, 36 m, 64 m and 100 m, respectively, as before.

A graph of these results looks like Figure 1.26.

Questions

18 Draw a velocity–time graph and a displacement–time graph for a body that is at rest and then accelerates at $3\,\mathrm{m\,s^{-1}}$ for 5 seconds.

19 Draw a velocity–time graph and a displacement–time graph for a car that is travelling at a uniform velocity of $8\,\mathrm{m\,s^{-1}}$ for 10 seconds.

20 Draw a velocity–time graph and a displacement–time graph for an object that starts at $12\,\mathrm{m\,s^{-1}}$ and gradually slows down and stops in 15 seconds.

Acceleration–time graphs

As the title indicates these are graphs which plot the acceleration of an object with time. In this section we will only consider examples of graphs where the acceleration is constant. Although it is possible to deal with situations where the acceleration increases or decreases, we will only consider uniform acceleration here.

Consider an object which goes from rest to $30\,\mathrm{m\,s^{-1}}$ in 20 seconds. If asked to draw the velocity–time graph

for this object using the method shown previously, you would end up with a graph like that shown in Figure 1.27.

The acceleration of this object is calculated using
$a = \frac{v - u}{t} = (30 - 0)/20 = 1.5\,\mathrm{m\,s^{-2}}$.

Using the graph we could calculate the acceleration at 5 s.

$a = \frac{v - u}{t} = (7.5 - 0)/5 = 1.5\,\mathrm{m\,s^{-2}}$

No matter at what time we calculate the acceleration, the result is always $1.5\,\mathrm{m\,s^{-2}}$. Our acceleration–time graph is therefore a horizontal straight line at $1.5\,\mathrm{m\,s^{-2}}$.

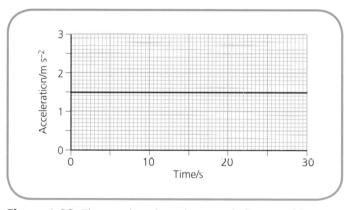

Figure 1.28 The acceleration–time graph for our object

This is the shape of an acceleration–time graph for an object with a steadily increasing speed or uniform acceleration.

Consider another example of an object which goes from rest to $40\,\mathrm{m\,s^{-1}}$ in 8 seconds.

If asked to draw the velocity–time graph for this object using the method shown previously, you should end up with a graph like that shown in Figure 1.29 overleaf.

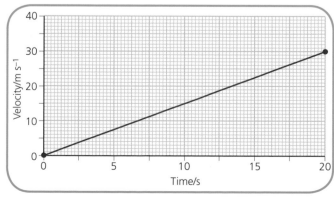

Figure 1.27 A velocity–time graph

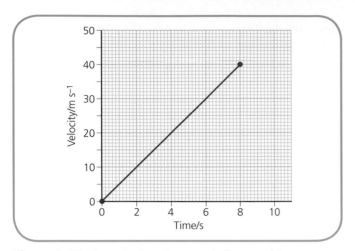

Figure 1.29 The velocity–time graph for our object

The acceleration of this object is calculated using
$a = \dfrac{v - u}{t} = (40 - 0)/8 = 5\,\mathrm{m\,s^{-2}}$.

Using the graph we could calculate the acceleration at 4 s.

$a = \dfrac{v - u}{t} = (20 - 0)/4 = 5\,\mathrm{m\,s^{-2}}$

It does not matter at what time we calculate the acceleration, the result is always $5\,\mathrm{m\,s^{-2}}$. This gives us a graph of a horizontal straight line at $5\,\mathrm{m\,s^{-2}}$ as shown in Figure 1.30.

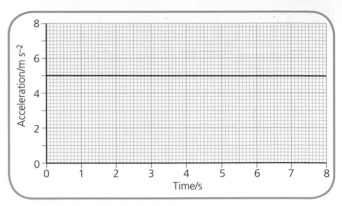

Figure 1.30 The acceleration–time graph for our object

This is the same type of graph as the previous example but the acceleration is greater and therefore the slope of the velocity–time graph is greater. The slope/gradient of the velocity–time graph is the acceleration!

Making the connection

In our analysis of graphs so far we have considered the link or relationship between displacement–, velocity– and acceleration–time graphs. You should be able to draw the displacement–time graph from a velocity–time graph and the acceleration–time graph from a velocity–time graph.

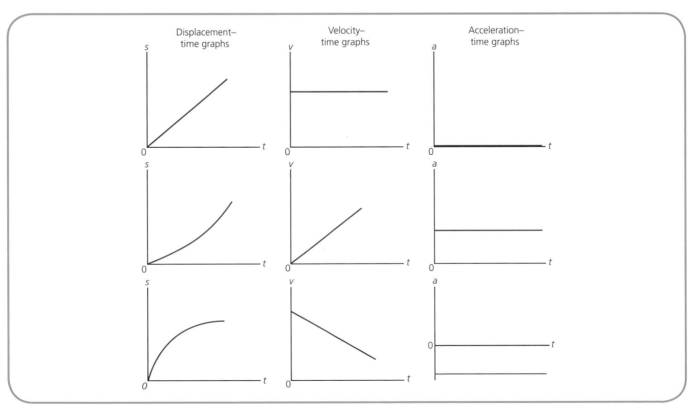

Figure 1.31 Making connections between the different graphs

Objects moving under the Earth's gravitational field

From earlier work you should be aware that objects falling due to the gravitational attraction of Earth will accelerate downwards towards the centre of the Earth.

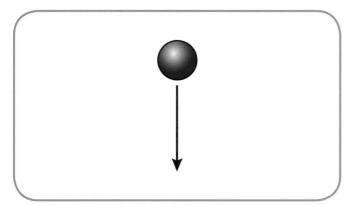

Figure 1.32 The ball drops to Earth as a result of gravity

Measurement of the acceleration due to gravity

A simple way to do this is to drop a ball from a height of around 2.0 m and time how long it takes for the ball to hit the ground.

A student carried out this investigation and repeated the experiment five times. Some possible results are given here:

0.65, 0.68, 0.62, 0.7, 0.68 s.

This gives a mean value of 0.67 s for a ball to fall a displacement of 2.0 m.

Using the equation $s = ut + \frac{1}{2}at^2$ (and removing the ut term as the ball is dropped from rest) we have

$$s = \frac{1}{2}at^2.$$

This can be rearranged to give $2s = at^2$ which then simplifies to $\frac{2s}{t^2} = a$.

This gives $\frac{4.0}{0.67^2} = a = 8.9\,\mathrm{m\,s^{-2}}$.

This is a reasonable approximation as the accepted value for the acceleration due to gravity (g) is $9.8\,\mathrm{m\,s^{-2}}$. You may have used the figure of $g = 10\,\mathrm{m\,s^{-2}}$ in previous studies, but the accepted value for this course is $9.8\,\mathrm{m\,s^{-2}}$. This means that an object dropped will accelerate at $9.8\,\mathrm{m\,s^{-2}}$ unless acted on by another force. As a result, the velocity–time graph for a dropped object will look like Figure 1.33.

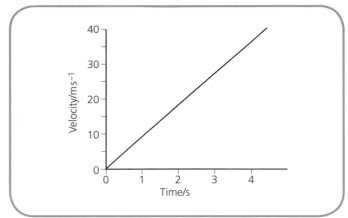

Figure 1.33 The velocity–time graph of a dropped object

The velocity–time graph of a falling object will be a straight-line graph with a slope of $9.8\,\mathrm{m\,s^{-2}}$, starting at rest and increasing its velocity.

This idea can be extended slightly to allow for objects which are thrown up and then return to Earth or objects which fall to Earth and then rebound. When we take into account objects moving up and down in opposite directions, and given that velocity is a vector quantity, we need to assign a positive or negative value to indicate which direction the object is travelling. For a dropping object, this is relatively straightforward.

A ball is released and falls towards the Earth. Its velocity–time graph is of an object gradually accelerating. Its velocity increases at a constant rate and therefore the shape of the graph is a straight line of slope 9.8.

For an object thrown upwards and allowed to fall back to its original position, it is slightly more challenging (see Figure 1.34).

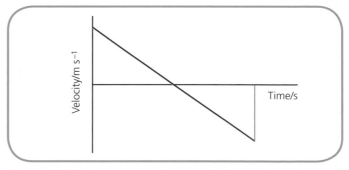

Figure 1.34 The velocity–time graph of an object thrown upwards and falling back to its original position

Schools have many devices now which can plot the motion of an object under certain circumstances. A simple experiment is to hold a motion sensor above the ground at a certain height and drop a large object like a football or basketball. The motion sensor sends small ultrasonic signals which travel out, reflect from an object and are then detected by the sensor. These reflected signals allow the sensor to measure displacement, velocity and acceleration. In this experiment the sensor should be set to measure velocity with respect to time.

Suppose a ball is released and allowed to hit the ground, rebound, rise and fall again. The detector plots its velocity and the graph produced will be similar to that shown in Figure 1.35.

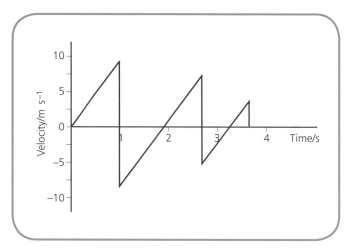

Figure 1.35 The velocity–time graph of a bouncing object

Explanation

As the ball falls it increases its velocity which is in the downward direction. It hits the ground and, very quickly, changes direction and rebounds upwards. This velocity is in the opposite direction to the original velocity. As it rises, its velocity decreases until it slows down and stops for a brief time at the top of its bounce.

It then falls downward, increasing in downwards velocity until it hits the ground again. It does not then rebound to the same height as some of its energy is lost when it hits the ground.

This sort of graph can appear difficult initially as it is easy to confuse the slopes of the graph with the direction the ball is travelling in, but that is not the way to interpret the graph.

- When a ball is dropped it accelerates and increases its velocity.

- When it hits the ground, it stops for a brief time and then rebounds upwards. This direction is opposite to its original direction and therefore its velocity has an opposite sign.

- The moment the ball leaves the ground it starts to slow down as gravity 'acts' in the downward direction.

- The ball heads upwards while accelerating in a downwards direction. This is difficult to comprehend at first but a way of considering it is to accept that all objects acted on by gravity will be accelerated downwards. It may be travelling upwards but it is slowing down. Gravity is causing it to decelerate.

- It reaches its highest point, where its velocity is zero for an instant (this is when it cuts the x-axis) and then it starts to accelerate downwards and 'repeats' its downwards journey.

- This continues until the ball stops bouncing.

The previous section considered objects falling under the effect of gravity in a vertical direction. You should also be aware that objects accelerate down slopes. Most of you would also accept that the acceleration down the slope increases with the angle of the slope.

Activities

Set up a track and a trolley on a small slope. Release the trolley and measure its acceleration. This could be done by the use of light gates placed at the side of the slope. Increase the slope slightly and measure the acceleration again. Repeat this for a range of angles.

Table 1.1 gives some possible results.

Angle of slope, $\theta/°$	Acceleration, $a/\mathrm{m\,s^{-2}}$
0	0.0
5	0.9
10	1.7
15	2.5
20	3.4

Table 1.1

Questions

- Does the acceleration increase as the angle of the slope increases?
- Is there a mathematical relationship?
- Can we manipulate or combine the results to obtain a relationship that will allow us to predict the acceleration at, for example, 25 or 35°?

A simple test is to divide one number by the other and look for a constant value.

Copy Table 1.1 and add two additional columns.

Complete the third column with $\frac{a}{\theta}$ and insert the answer into the appropriate column. Is there a pattern or a consistent set of results?

Complete the fourth column with $\frac{a}{\sin\theta}$.

The explanation of why these results are typical is examined later in the text.

Worked Example 1.7

A ball is thrown vertically upwards from a balcony at 4.5 ms⁻¹. It reaches a maximum height and then falls back to the ground 3.2 m below the level at which it was released.

a) Calculate the maximum height above the balcony reached by the ball.

When the ball reaches its highest point, its velocity is 0 ms⁻¹.

$v = 0, u = +4.5, a = -9.8, t = ?, s = ?$

To calculate the height(s) we can use $v^2 = u^2 + 2as$

$$0^2 = (4.5)^2 + 2 \times (-9.8)s$$

$0 = 20.25 - 19.6s$

$s = 1.03$

$= 1.0$ m (above balcony)

b) How long was it in the air before it struck the ground?

$v = ?, u = +4.5, a = -9.8, t = ?, s = -3.2$

When the ball strikes the ground it lands 3.2 m below the level of the balcony. Accordingly, the final displacement is −3.2 m.

To solve for t we could use $s = ut + \frac{1}{2}at^2$ but this would involve solving a quadratic equation which may not have a simple solution.

→

Alternatively, it can be done by a two-step process.

The first involves calculating v using $v^2 = u^2 + 2as$.

Substituting gives: $v^2 = 20.25 + 2 \times (-9.8) \times (-3.2)$

$= 82.97$

$v = -9.1$ (negative chosen as velocity is downwards)

Now that we have $v = -9.1$ ms^{-1} we can use $v = u + at$.

$-9.1 = 4.5 - 9.8t$

$t = 1.387...$

$= 1.4$ s

Consolidation Questions

1 Explain the difference between a scalar and a vector.

2 List two scalars and two vectors.

3 A runner runs 3 km due North then 4 km due East. He takes 30 minutes to run the first leg and another 30 minutes to run the second leg.

Calculate the average speed and average velocity of the runner and give the direction in degrees.

4 A car is driving along a road at 15 ms^{-1} when the driver is forced to brake heavily. The car stops in a distance of 38 m.

Calculate the acceleration of the car.

5 A hot air balloon of mass 200 kg is descending vertically at 4.0 ms^{-1}. A 10 kg bag of sand is released and it hits the ground 6.0 s later. From what height was it released?

6 a) Sketch the velocity–time graph for a vehicle that carries out the following journey: the vehicle starts at 3.0 ms^{-1} and maintains it for 10 seconds, then accelerates to 9.0 ms^{-1} in 12 seconds; it then slows down and stops in a further 6.0 seconds.

 b) Sketch the acceleration–time graph for this velocity–time graph.

7 A ball is dropped from a height of 1.9 m. It rebounds from the ground with a velocity of 5.0 ms^{-1}.

 a) Calculate the velocity at which it hits the ground.

 b) How long does it take to reach the ground?

 c) To what height does it rebound?

 d) Calculate the time taken to reach its maximum rebound height.

 e) Draw a velocity–time graph for its motion.

8 A stone is thrown upwards with a velocity of 15 ms^{-1}.

 a) Calculate the time taken for the stone to reach the highest point.

 b) What is its velocity at this point?

 It then falls back down to its original position.

 c) Sketch a velocity–time graph for its motion.

 d) Sketch an acceleration–time graph for its motion.

9 During a flight an aircraft is travelling with a velocity of 120 ms^{-1} due North (000°). There is a wind of velocity 25 ms^{-1} at a bearing of 055°.

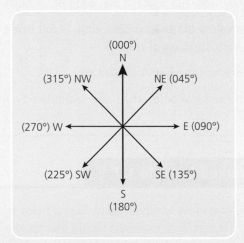

Figure 1.36

Calculate the magnitude and resultant velocity of the aircraft.

10 The manufacturers of tennis balls require that the balls meet a given standard. When dropped from a certain height onto a test surface, the balls must rebound to within a limited range of heights.

The ideal ball is one which, when dropped from rest from a height of 3.75 m, rebounds to a height of 2.15 m, as shown in Figure 1.37.

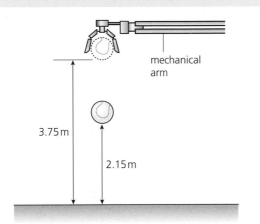

Figure 1.37

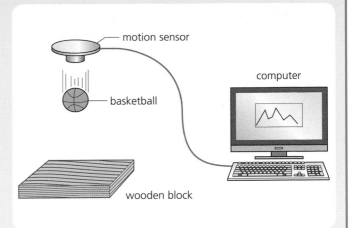

Figure 1.39

A displacement–time graph for the motion of the basketball from the instant of its release is shown in Figure 1.40.

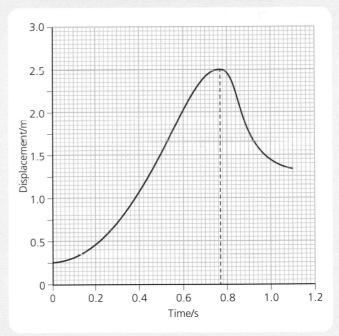

Figure 1.40

Assuming air resistance is negligible, calculate

a) the speed of an ideal ball just before contact with the ground

b) the speed of this ball just after contact with the ground.

11 A spectator at A walks to C, the opposite corner of a playing field, by walking from A to B and then from B to C as shown in the diagram below.

The distance from A to B is 75 m. The distance from B to C is 215 m.

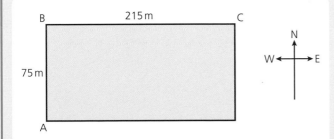

Figure 1.38

By scale drawing or otherwise, find the resultant displacement. Magnitude and direction are required.

12 A basketball is held below a motion sensor. The basketball is released from rest and falls onto a wooden block. The motion sensor is connected to a computer so that graphs of the motion of the basketball can be displayed (see Figure 1.39).

a) i) What is the distance between the motion sensor and the top of the basketball when it is released?

ii) How far does the basketball fall before it hits the wooden block?

iii) Show, by calculation, that the acceleration of the basketball as it falls is 7.6 m s^{-2}.

b) The wooden block is replaced by a block of sponge of the same dimensions. The experiment is repeated and a new graph obtained. Describe and explain two differences between this graph and the original graph.

13 A 2.5 kg mass slides down a frictionless slope as shown.

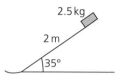

Figure 1.41

a) It accelerates from rest, from the top of the slope, down the slope at 5.6 m s⁻².

i) Calculate the time it takes to travel 1 m down the slope and the time it takes to reach the end of the slope.

ii) Calculate its velocity at 1 m and 2 m.

iii) Calculate its average velocity between the 1 m and 2 m points. Is this value the same as you would calculate for the mass when it passes the 1.5 m point? Justify your answer.

14 In a car crash test a vehicle is accelerated from rest to 34 m s⁻¹ in 8.8 s. It strikes another vehicle and they combine and eventually stop in a distance of 9.6 m.

a) i) Calculate the initial acceleration of the test vehicle.

ii) Calculate the acceleration of the combined vehicles as they slowed down.

2 Forces, energy and power

The work of Galileo and Newton explained correctly the motion of an object when acted on by a force or combination of forces. In this chapter we will look at the relationship between the force(s) acting on an object and its motion. For the sake of simplicity we will consider motion in a straight line only.

From earlier work we know that an unbalanced force causes an object to accelerate. If a force is applied to an object and no other forces act, this object will continue to accelerate steadily and increase its velocity for as long as the force acts. (This rule only holds until the velocity of the object approaches the speed of light at which point other factors apply.)

Why is it then that in normal life we rarely approach high velocities? The main determining factor is **friction**. Whether it be air or water or an object in direct contact with the moving body, the friction caused by these 'contacts' will reduce the unbalanced force acting.

For example, when a car starts, the accelerator is pressed and the engine applies a force. This causes the car to accelerate and increase its velocity. Eventually it reaches its final (maximum) velocity even though the accelerator pedal is still being pressed. The engine has to continue applying a force to enable the car to remain at its set velocity. It cannot increase its velocity forever.

The force required to push the car through the air at high speed (from the engine) is balanced by the road friction and drag.

At the point where the forward force of the engine is equal to the drag and road friction acting in the opposite direction, the overall force on the car is zero. If there is no unbalanced force acting on the car then its velocity does not change. It remains at whatever velocity it had reached when the forces became 'balanced'.

In all of the situations given in Figure 2.1, the forces are balanced.

There are many everyday situations which require an understanding of balanced and unbalanced forces to be understood fully. The next section discusses the physics behind just a few of them.

Consider a raindrop. The drop forms in the upper atmosphere and when conditions are correct it falls to the ground. It falls from a great height but it does not hit us or land on the ground with a very high velocity. Why?

The raindrop falls because of gravitational attraction. This force causes the drop to accelerate. As it falls and accelerates, the air friction affecting the drop increases. The drop eventually reaches a velocity when the force attracting it (its weight) is equal to the frictional forces opposing its motion. At this point it does not increase its velocity; the velocity remains constant. This is referred to as its **terminal velocity**.

A skydiver exits from an aeroplane and falls for a few minutes before opening her parachute.

Figure 2.1 Examples of situations where the forces are balanced

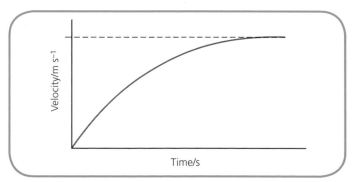

Figure 2.2 The velocity–time graph for a skydiver

How do we explain in terms of forces acting on the skydiver the shape of the velocity–time graph?

As the skydiver leaves the aeroplane, her weight is acting downwards and she accelerates downwards. As her velocity downwards increases, so does her drag (air friction) and this reduces the unbalanced force downwards. She still accelerates downwards but at a lesser rate. She ultimately reaches a velocity which is so great that the drag equals her weight and there is no resultant force downwards. She stays at the same velocity but does not accelerate.

How could we explain why a small stone has a greater terminal velocity than a small feather?

Terminal velocity occurs when the force downwards (weight) is balanced by the drag force. A feather has a very small weight and a large surface area so the velocity it needs to reach for the drag and weight to be balanced is much smaller than that of a stone, which has a much greater weight and similar surface area.

Why do young children (and many others) believe that heavy things fall more quickly than light things?

There could be many reasons for this, including that this is what their parents tell them. Children may see lightweight things such as leaves, paper and balloons falling slowly and see bricks, stones and solid objects falling more quickly. This is a reasonable connection to make when young. Children will not appreciate the other forces acting on the objects.

Applying the principle of forces acting on objects and their motion

The principle of forces acting on objects and their motion can be applied to many everyday situations.

Vehicles

The top speed of a car or truck is determined by a number of factors. The power and gearing of the engine, the weight and the aerodynamic shape of the vehicle all affect its final velocity. Table 2.1 gives some data on the top speed of various vehicles.

Golf

The driving distances of modern golfers greatly exceed those of champions of only 20 years ago. All the increases are due to improvements in golf technology. Larger club heads, longer shafts and the aerodynamic shape of the ball have all added to the distance generated.

The distance travelled for similar balls is due to the angle and velocity with which the ball leaves the club and the effect air resistance has on the ball. For example, a ball leaves a club with a velocity of $63\,\mathrm{m\,s^{-1}}$ and the instant it leaves the club it begins to slow down dramatically. We do not 'see' this drop in speed as the initial velocities are still large and it is difficult to appreciate.

Some general velocities of ball and club during a drive are detailed below:

Club head speed $= 45\,\mathrm{m\,s^{-1}}$
Ball speed immediately after impact $= 63\,\mathrm{m\,s^{-1}}$
Landing speed of ball $= 20\,\mathrm{m\,s^{-1}}$

Rocket take off

In order for a rocket to take off, the force upwards (thrust) must be greater than any downward force (weight). When the thrust exceeds the weight, there is a net unbalanced force. This force causes the rocket to accelerate upwards.

In general, the thrust of an engine remains fairly constant during lift off and some large rockets have a

Make of car	Power/BHP	Mass/kg	Acceleration (0−60 mph)/s	Top speed/mph
Audi TT	246	1430	5.5	155
Mini Clubman	93	1395	11.3	114
Land Rover Discovery	188	2419	11	112
Porsche Cayman 2.7	245	1360	5.9	160
Nissan Leaf	147	1557	8.3	90

Table 2.1 Table of cars and their associated performance data

Figure 2.3 A modern rocket delivery system

thrust capacity which is slightly less than the weight of the rocket.

However within a few seconds some fuel has been burned and ejected so the thrust is greater than the weight and the rocket lifts off with a small acceleration.

As more fuel is burned and ejected, the weight and mass of the rocket reduces and the acceleration increases. Within a short time the acceleration and velocity of the rocket are very high.

Objects being raised or lowered using ropes and wires

Cranes, elevators and hoists raise and lower large objects using a variety of pulleys and cables but they all operate on the general principle of balanced and unbalanced forces.

Consider the example shown in Figure 2.4.

Let us say that the weight of the beam is 24 000 N. As the crane begins to raise the beam, the cable increases its upwards pull until this equals the weight of the beam. The cables are being raised by the crane's

Figure 2.4

engine and this leads to a stretching of the cable. This 'stretching' transfers the force to the beam causing it to be raised.

The stretching of the cable is often referred to as **tension**. Steel cables are made of many thin steel wires which are held together tightly. These cables have great tensile strength and are used to support bridges and other structures.

Figure 2.5 Modern bridges are supported by strong steel cables

When the tension is greater than the weight of the beam, there is an unbalanced force in an upwards direction and the beam accelerates upwards.

24 500 N ↑

24 000 N ↓

Once the beam is moving at a steady velocity the tension is reduced so that the tension matches the weight. At this point, the forces are balanced.

24 000 N ↑

24 000 N ↓

As it reaches the top of its lift, the tension is reduced and this leaves a resultant force downwards; the beam slows down and stops.

23 500 N ↑

24 000 N ↓

At this point in mid-air the forces are balanced with the weight of the beam equal to the tension in the cable.

24 000 N ↑

24 000 N ↓

The same principle applies in the fairground game which involves people running along a slippery channel while attached to a large, elastic bungee rope. As they run, their movement stretches the rope creating a tensile force in the opposite direction to their running. At a certain point, the force of the person's feet as they run or push in one direction is balanced by the tension in the rope. This only lasts for a few seconds and when the person's feet slip, the tension becomes the unbalanced force and the person is accelerated backwards along the channel.

Description of a combined lift system

Many older elevator systems used to operate in tandem to reduce the energy required to raise and lower people or goods. In such a system, two carriages are connected by a steel cable which is wound round a large wheel. As one carriage is raised, the other carriage is lowered.

The force required to raise and lower the carriages is to overcome the friction in the system and the weight of any additional passengers, not to overcome the weight of the carriages.

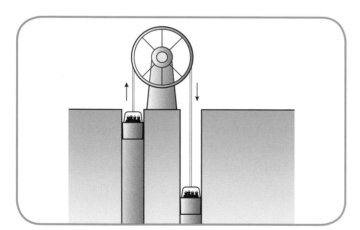

Figure 2.6

Figure 2.7 The Falkirk wheel

A similar system exists in the Falkirk wheel. The two bays counterbalance each other and additional force is needed only to overcome friction. It is remarkably efficient.

Forces in two dimensions

In the previous section, we have mainly been considering the effects of forces on motion, and combinations of such, in one dimension. In other words, the forces and motion act in the same or opposite directions. We can analyse systems when the forces act in two dimensions by utilising the properties of vectors. We can split vectors into various components or directions dependent upon the situation being considered.

The following illustrates this idea with a simple example and goes back to our work with displacement from the previous chapter.

A person travels as shown in Figure 2.8.

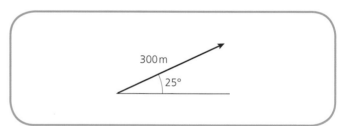

Figure 2.8

This displacement can be made up of two other vectors, one 'horizontal' and one 'vertical'.

These are the horizontal and vertical components of the original vector and are shown in Figure 2.9.

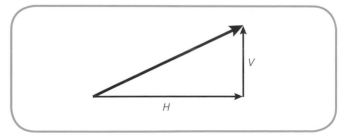

Figure 2.9

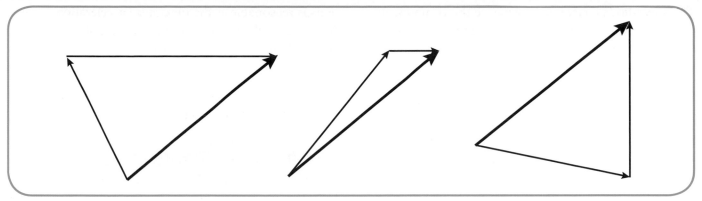

Figure 2.10 Different possible vector routes

In this case we separated the vector into horizontal and vertical components. We could separate a vector into other components such as those shown in Figure 2.10.

All of these 'components' are valid ways of separating the original vector but for most cases horizontal and vertical components are the easiest and most useful to deal with.

Resolving vectors

Vectors can be split into horizontal and vertical components by a number of methods.

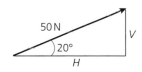

Algebraically

We can use trigonometrical relationships to calculate the magnitude of the horizontal and vertical components.

$$\frac{\text{horizontal}}{\text{hypoteneuse}} = \cos\theta$$

Therefore, $H = \text{hypoteneuse} \times \cos\theta$

hypoteneuse = 50 N

$H = 50\cos 20° = 47.0$

$$\frac{\text{vertical}}{\text{hypoteneuse}} = \sin\theta$$

Therefore, $V = \text{hypoteneuse} \times \sin\theta = 50\sin 20° = 17.1$

A check could be made using Pythagoras' theorem.

$$H^2 + V^2 = \text{hypoteneuse}^2$$

$$50^2 = 47.0^2 + 17.1^2$$

Graphically

Using a protractor and suitable scale we can calculate the horizontal and vertical components.

- Use 10 N = 1 cm.

- Measure 20° with a protractor and draw a line. Mark a point (A) and measure 5 cm. Mark another point (B).

- Using the protractor draw a horizontal line from A. Draw a vertical line down from B.

- Measure the lengths of these lines and using the scale calculate the horizontal and vertical components.

Worked Example 2.1

Skiers can be aided by a hoist that pulls a spring-loaded wire, which then pulls skiers along the snow.

Figure 2.11

The wire applies a steady force to the skier. The tension in the wire is a steady 150 N at an angle of 45°. Calculate the horizontal component of the force pulling the skier.

$H = 150\cos\theta = 150\cos 45° = 106.1$ N

These methods enable us to separate the components of vectors which allows calculations of how forces affect the motion of objects in a range of quite complex situations.

Motion on a slope

An object on a horizontal surface is at rest. The weight of the object acts in a vertical direction. As such the object does not move. If the surface is raised at one end, the object begins to slide down. The higher we raise the surface, the more quickly the object slides. The weight of the object does not change but the component of weight acting down the slope increases and this increases the motion of the object.

The weight of the object is calculated using

$$W = mg$$

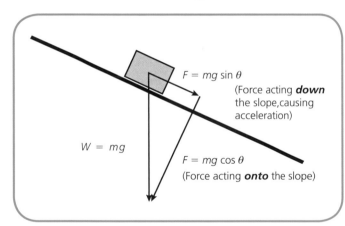

Figure 2.12 Components of weight acting on an object

The weight of the object (mg) can be considered to have two components:

1 One component of that, $mg \sin \theta$, causes the object's motion. It makes it accelerate down the slope.

2 The other component acts in a perpendicular direction onto the surface of the slope.

The component of the weight acting down the slope is given by

$$F = mg \sin \theta$$

As acceleration $a = \dfrac{F}{m} = \dfrac{mg \sin \theta}{m} = g \sin \theta$

This indicates that the acceleration of the object varies with the sine of the angle. Compare this with the results given on page 19 and check whether it tallies.

This is reasonable as the maximum acceleration is $9.8\,\text{m s}^{-2}$ which would occur when the slope is perpendicular. When perpendicular $\theta = 90°$, $\sin \theta = 1$ and $mg \sin \theta = 9.8\,\text{m s}^{-2}$.

The acceleration cannot be greater than $9.8\,\text{m s}^{-2}$ as that is what happens when the object falls freely.

Worked Example 2.2

This worked example ignores the effect of friction.

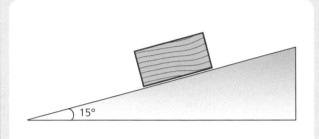

Figure 2.13

A 2 kg block is placed on a slope as shown and released.

a) Calculate the acceleration of the block.

$$a = \frac{F}{m} = \frac{mg \sin \theta}{m} = g \sin \theta = 9.8 \times \sin 15°$$

$$a = 2.5\,\text{m s}^{-2}$$

b) Calculate the force acting onto the slope.

$$F = mg \cos \theta = 2 \times 9.8 \times \cos 15° = 18.9 = 19\,\text{N}$$

The block is replaced by one of a mass of 10 kg.

c) Calculate the acceleration of the block.

$$a = \frac{F}{m} = \frac{mg \sin \theta}{m} = g \sin \theta$$

$$a = 2.5\,\text{m s}^{-2}$$

d) Calculate the force acting onto the slope.

$$F = mg \cos \theta = 10 \times 9.8 \times \cos 15 = 94.7 = 95\,\text{N}$$

These calculations show that the acceleration is dependent upon the angle of the slope and not the mass of the object. This is paralleled in some ways by the dropping of two masses of different size. Under the action of gravity the masses fall to the Earth at the same rate. The mass is not the determining factor if friction is not included.

Questions

1   A 4.0 kg trolley is released on a slope at an angle of 15°. (Ignore friction in your calculations.)

   a) Calculate the component of weight acting down the slope.

   b) Calculate the acceleration of the trolley.

2   A 2.0 kg trolley is released on a slope and accelerates at $3.0\,\mathrm{m\,s^{-2}}$ (ignore friction).

   a) Calculate the unbalanced force on the trolley.

   b) What angle does the slope make with the horizontal?

   c) What difference in acceleration would there be if we replaced the trolley with a similar one of mass 1.0 kg?

3   In a class experiment an egg is placed in a trolley on a slope as shown in Figure 2.14.

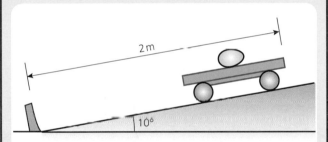

Figure 2.14

The trolley is released and collides with the barrier at the bottom of the slope.

   a) Calculate the acceleration of the trolley.

   b) What velocity does it have when it strikes the barrier?

Friction

We can adapt the situation to include the effect of friction on the system.

Example 1

A train of mass 22 000 kg and engine force 12 000 N travels along a railway track. It is travelling at a constant velocity.

The track now rises at an angle of 1°.

Calculate the component of the weight acting down the slope.

$F = mg \sin \theta = 22\,000 \times 9.8 \times \sin 1° = 3800\,\mathrm{N}$

What must the engine force increase to in order for the train to remain at a constant velocity?

It must increase its engine force to $12\,000 + 3800 = 15\,800\,\mathrm{N}$. A slope of 1° means the train must increase its power by about 32% to remain at the same velocity.

Example 2

A children's slide is shown in Figure 2.15.

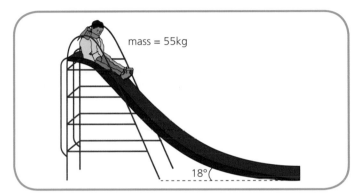

Figure 2.15

The slide is 8 m long.

Calculate the acceleration of the child and his velocity at the bottom of the slide.

$$F = mg \sin \theta \text{ and } a = \frac{F}{m} = \frac{mg \sin \theta}{m}$$

$$= \frac{55 \times 9.8 \times \sin 18°}{55}$$

$$= 9.8 \times \sin 18° = 3.0\,\mathrm{m\,s^{-2}}$$

To calculate the velocity of the child at the bottom we use the equations of motion.

We have the acceleration, initial velocity and displacement.

$$v^2 = u^2 + 2as = 0^2 + 2 \times 3.0 \times 8 = 48$$

$$v = 6.92\ldots\,\mathrm{m\,s^{-1}}$$

$$= 6.9\,\mathrm{m\,s^{-1}}$$

This is not correct as the slide would have to be frictionless and be at the same angle for the duration of the slide but it does give us an approximation for the velocity at the bottom. The reality is that it will be slower due to frictional forces.

The principle of resolving forces into components and then analysing the effects of those forces can also be applied to situations where we combine forces. As forces are vectors, we use the rules of combining vectors to calculate what the combined or total resultant will be.

Combining vectors

Vectors can be combined algebraically and graphically in a similar way to that described earlier on page 27.

The vectors are combined by joining one vector with another as shown in Figure 2.16.

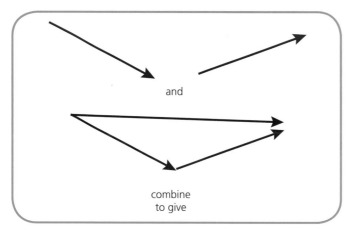

and

combine
to give

Figure 2.16 Combining vectors

This 'new' vector is the **resultant vector**. This vector has the equivalent effect or result of combining the two vectors. The resultant vector has the same magnitude and direction regardless of the order in which we combine them.

For example:  combined with

can be represented by following one with the other such as:

This has the same resultant as:

The resultant vector in both cases is exactly the same:

A common mistake made by students is the translation of the forces acting on an object into an appropriate vector diagram. It is essential the vectors are combined 'nose to tail'.

For example, when dealing with a situation in which two forces act on an object, a common error is to take the force diagram and simply draw a line across from the vectors. This is incorrect.

The correct vector diagrams which give the resultant are:

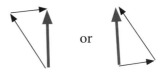

or

Questions

4 Draw a simple sketch which shows the resultant vector in the following cases:

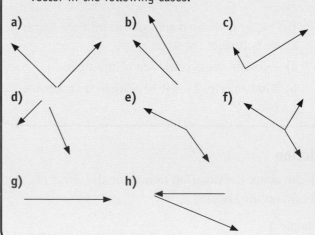

a) b) c)

d) e) f)

g) h)

Having drawn the correct vector diagram, the next stage is to determine the magnitude and direction of the resultant in one of two ways:

1 Algebraically

This force diagram becomes a vector diagram.

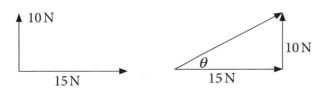

Calculating the resultant magnitude in this case is straightforward using Pythagoras' theorem.

$$10^2 + 15^2 = 100 + 225 = 325$$

The square root of $325 = 18$

The angle can be calculated using $\tan \theta = 10/15 = 0.67$

Therefore $\theta = 33.8°$

2 Graphically

Choose a suitable scale: $1\,\text{cm} = 5\,\text{N}$ for example. Using a ruler draw the shape to scale, draw the hypotenuse and then measure the distance. Convert this distance on the appropriate scale and this gives the magnitude of the resultant. Using a protractor, measure the angle.

The same principle can be applied to more complex situations where the forces are not at right angles to each other.

Worked Example 2.3

Calculate the resultant force in this example.

From the diagram combine the forces in vector fashion.

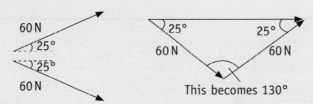

This becomes 130°

Using the angles given in the diagram we can draw the vector diagram and calculate the included angle.

Knowing both vectors and the included angle allows us to calculate the magnitude of the resultant using the cosine rule, or possibly the sine rule if we calculate one of the other angles.

Using the cosine rule:

$$a^2 = b^2 + c^2 - 2bc \cos A$$

$$a^2 = 60^2 + 60^2 - 2 \times 60 \times 60 \times \cos 130°$$
$$= 3600 + 3600 - 7200 \times -0.64$$

$$a^2 = 11828$$

$$a = 108.8\,\text{N} = 110\,\text{N}$$

As the forces have the same angle and one acts 'upwards' and one 'downwards', the vertical components effectively cancel each other out leaving a resultant vector which acts horizontally.

Another way would be to split each vector into its horizontal and vertical components and simply combine all the components.

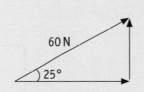

$$H = 60 \cos 25° - 60 \times 0.91$$
$$= 54.4\,\text{N}$$

$$V = 60 \sin 25° = 60 \times 0.42$$
$$= 25.4\,\text{N}$$

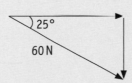

$$H = 60 \cos 25° = 60 \times 0.91$$
$$= 54.4\,\text{N}$$

$$V = 60 \sin 25° = 60 \times 0.42$$
$$= 25.4\,\text{N}$$

Resultant horizontal =
$54.4\,\text{N} + 54.4\,\text{N} = 108.8\,\text{N}$

Resultant vertical =
$25.4\,\text{N} + -25.4\,\text{N} = 0\,\text{N}$

Worked Example 2.4

Calculate the magnitude and direction of the resultant of this combination of vectors.

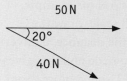

50 N

20°

40 N

As one of the forces acts only in the horizontal direction, it is straightforward to deal with. A possible solution here is to take the 40 N vector and split it into horizontal and vertical components. We then combine both the horizontal vectors to give the total horizontal component. This is combined with the vertical component to give the resultant.

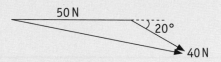

50 N
20°
40 N

$H = 40 \cos 20° = 37.6 \text{ N}$
$V = 40 \sin 20° = 13.7 \text{ N}$

This gives us a vector diagram of

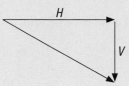 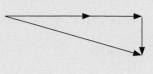

H

V

The resultant can be calculated using Pythagoras:
$a^2 + b^2 = c^2$

$(50 + 37.6)^2 + 13.7^2 = c^2 = 7861.5$

$c = 88.7 = 89 \text{ N}$

The angle can be calculated using simple trigonometry:

$\tan \theta = \frac{13.7}{87.6} = 0.156$

Therefore $\theta = 8.9°$

Questions

5   Calculate the magnitude and direction of the resultant vector in the following examples.

a)

6 N

10 N

b)

20 N

15 N

c)

4 N

12 N

d)

5 N

10 N   3 N

e)

100 N

200 N

50 N

f)

10 N

45°

10 N

g)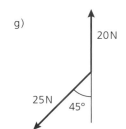

20 N

25 N

45°

h)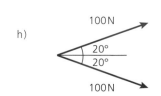

100 N

20°
20°

100 N

Figure 2.17

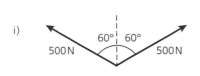

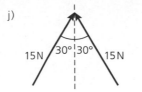

Figure 2.17 continued

6 A boat crosses a river at $3\,\text{ms}^{-1}$ as shown in Figure 2.18. The current in the river is a constant $1.5\,\text{ms}^{-1}$.

   a) What is the resultant velocity of the boat?

   b) How far away from the pier would the boat land if no correcting action was taken?

7 An aeroplane is flying due East at a velocity of $210\,\text{ms}^{-1}$. It flies into a headwind of $30\,\text{ms}^{-1}$.

   a) What is its resultant velocity?

   b) The wind changes direction and blows due North. What is the new resultant velocity?

8 On transatlantic flights between the UK and North America, the average flight time is less in one direction than the other. Can you suggest why this is so?

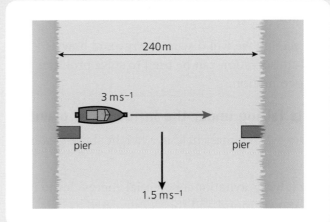

Figure 2.18

Motion, forces and energy

The study of motion and how forces affect that motion leads us to other physical concepts such as energy. How the concepts are related gives us an indication of the true nature of the physics concerned. It can help when carrying out a topic such as this to map out concepts, making it clear how the different parts are related. This can greatly help to clarify concepts in your mind and help you to organise the key underlying principles.

Energy

What is energy and how many different forms are there? This is a question many physics teachers dread. It is simple yet the explanations can become exceedingly complex. When Richard Feynmann (1918–1988), a Nobel Prize-winning physicist, was asked the question, 'What is energy?', he answered:

'There is a fact or, if you wish, a law governing all natural phenomena that are known to date. There is no known exception to this law – it is exact so far as we know. The law is called the conservation of energy.

It states that there is a certain quantity, which we call "energy", that does not change in the manifold changes that nature undergoes. That is a most abstract idea, because it is a mathematical principle; it says there is a numerical quantity which does not change when something happens.

It is not a description of a mechanism, or anything concrete; it is a strange fact that when we calculate some number and when we finish watching nature go through her tricks and calculate the number again, it is the same.

> It is important to realise that in physics today, we have no knowledge of what energy "is". It is not that way. It is an abstract thing in that it does not tell us the mechanism or the reason for the various formulas.'

Feynmann was a great physicist but also tried to make the subject popular and to improve the understanding of physics in general. Much has been written by and about him.

In this section we will discuss energy and how its overall conservation can be used to solve problems relating to motion.

Objects falling under the influence of gravity

Consider a 2.0 kg mass held at a height above ground of 3.0 m.

We say it has gravitational potential energy due to its height above ground and this is calculated by

$$E_p = m \times g \times h$$
$$= 2.0 \times 9.8 \times 3.0 = 58.8 \, \text{J}$$

If we release the ball, it will fall to the ground and its gravitational 'potential' energy will convert to kinetic energy. Therefore

$$E_k = \frac{1}{2} \times m \times v^2$$
$$= 58.8 \, \text{J}$$

This gives $v^2 = 2 \times \dfrac{58.8}{m} = 58.8$

$v = 7.67 = 7.7 \, \text{m s}^{-1}$

This question shows how we arrive at the same number before and after 'nature' does its work.

Let us consider it from an 'equation of motion' perspective. What velocity will it reach just before hitting the ground?

$v = ?$	$t = ?$
$u = 0$	$s = 3.0$
$g = 9.8 \, \text{m s}^{-2}$	

To calculate v we could use $v^2 = u^2 + 2as$
$$v^2 = 2 \times 9.8 \times 3.0 = 58.8$$
$$v = 7.7 \, \text{m s}^{-1}$$

These numerical results would seem to confirm the idea that energy has the same value before and after an event and that the form of energy has altered from one to another.

Worked Example 2.5

A mass of 4.0 kg is released from a height of 2.4 m. It falls and rebounds to a height of 1.8 m.

a) Calculate the potential energy of the object at each of the heights.

b) Calculate the kinetic energy of the object:
   i) just as it makes contact with the ground
   ii) just after it leaves the ground.

Why is there a difference in the kinetic energies of the object?

Can you account for this difference in energy calculated?

$E_p = mgh = 4.0 \times 9.8 \times 2.4 = 94 \, \text{J}$

This is transformed to kinetic energy of the mass prior to it hitting the ground.

$E_p = mgh = 4.0 \times 9.8 \times 1.8 = 71 \, \text{J}$

This is the kinetic energy the mass has as it rebounds.

$v_{\text{downwards}} = 6.9 \, \text{m s}^{-1}$

$v_{\text{upwards}} = 5.9 \, \text{m s}^{-1}$

The velocity upwards is less and this results in the mass not reaching the same height as it was dropped from. Our energy at the start does not equal our energy at the end of the rebound. There is an 'amount' of energy which appears to be missing. It is not missing: some energy was transformed to sound and heat as the mass struck the ground. This happens during almost every 'collision' and can explain why a bouncing ball will gradually lose more and more height as it bounces until it stops. Its initial energy is gradually transformed to sound and heat and is also 'lost' to friction as it moves through the air.

This next example considers motion on slopes where friction is acting.

A boy is on a sled at the top of a hill as shown in Figure 2.19.

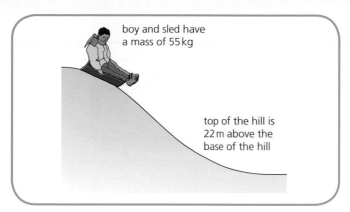

boy and sled have a mass of 55 kg

top of the hill is 22 m above the base of the hill

Figure 2.19

Calculate the potential energy of the boy at the top of the slope.

$$E_p = m \times g \times h = 55 \times 9.8 \times 22 = 11858 = 12000\,\text{J}$$

His velocity at the bottom of the slope was found to be $14.2\,\text{m s}^{-1}$.

Calculate the kinetic energy of the boy and sled at the bottom of the slope.

$$E_k = \tfrac{1}{2}mv^2 = \tfrac{1}{2} \times 55 \times 14.2^2 = 5545 = 5500\,\text{J}$$

There is a large difference in the energy of the boy at the bottom of the slope compared to the energy at the top of the slope. The energy is not 'missing'; it has been 'transformed' to another type.

In this case the energy has been converted to heat mainly due to friction as the sled moved down the slope. This is described in terms like the following:

- 'Energy has been used to overcome friction'
- 'Work has to be done to overcome friction and this results in a transfer of energy'
- 'Energy has been lost to heat'

All these explanations are reasonably valid and all refer to this idea that we have a certain amount of energy at the beginning of the process and we have the same amount at the completion. This principle of **energy conservation** is applicable to all situations.

The term 'work done' is used widely in these types of problems and we can apply it in some more detail to this example. As the boy slides down the slope, there is friction between his sled and the snow/ice. The component of his weight acting down the slope is greater than the friction force acting against him so he accelerates downwards. This friction force acts on him throughout his journey down the slope, however, and energy must be 'used' or transferred to overcome this.

Work is done in this situation. In doing this work, energy is transferred from the boy to the ice/sled. This results in an amount of energy being transformed from the 'amount' he had initially.

$$11858\,\text{J} - 5545\,\text{J} = 6313\,\text{J}$$

We can say that 6313 J of work was done in overcoming the friction.

From earlier work we know that

$$\text{work done} = \text{force} \times \text{displacement}$$

The length of the slope is 65 m. We can calculate the friction between the sled/ice using

work done = force × displacement

$6313 = \text{force} \times 65$

$\text{force} = 6313/65 = 97.1\,\text{N}$

The force of friction acting on the shoes is 97 N.

Worked Example 2.6

A truck of 3000 kg is travelling along a motorway at 24 m s^{-1}. The driver sees warning lights come on in the distance and brakes until the truck stops.

Calculate the kinetic energy of the truck when it is at 24 m s^{-1}.

$$E_k = \frac{1}{2} \times m \times v^2 = \frac{1}{2} \times 3000 \times 24^2 = 864\,000\,J$$

The truck stops in 72 m. Calculate the force of friction between the tyres and the road.

work done = force × displacement

$$864\,000 = \text{force} \times 72$$

$$\text{force} = 864\,000/72 = 12\,000\,N$$

What has happened to the kinetic energy? It has mainly been converted into sound and heat in the tyres, braking systems and tarmac on the road.

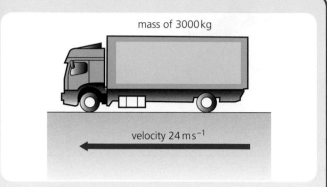
mass of 3000 kg

velocity 24 m s^{-1}

Figure 2.20

For Interest   The space shuttle landing – some surprising figures

Figure 2.21 A space shuttle

When the space shuttle begins its slow down and descend to land it has a mass of approximately 100 000 kg and is travelling at a velocity of approximately 13 000 m s^{-1}.

As it starts to make contact with particles in the upper atmosphere, its 'slow down' begins. Calculate its kinetic energy at this point.

(Note: This deals with only the kinetic energy of the shuttle and is only an approximation to give an indication of the energy involved.)

$$E_k = \frac{1}{2} \times m \times v^2$$

$$= 0.5 \times 100\,000 \times 13\,000^2$$

$$= 8.45 \times 10^{12}\,J$$

It comes to a halt 80 minutes later.

How much energy must it lose per second in order to come to a stop at the landing strip?

How do you think it does this?

At what rate must it lose energy on its descent to do this?

$$= 8.45 \times 10^{12}/(80 \times 60) = 1.76 \times 10^9\,J\,s^{-1} = 1.76\,GW!!$$

Questions

9 In the downhill race at the Winter Olympics, a skier drops a height of 900 m from the start to the finishing line. He covers a distance of 3.4 km. His winning time is 1 min 50 seconds and the speed at which he passes the finishing line is $35\,\text{m s}^{-1}$. The skier has a mass of 92 kg.

  a) Calculate the potential energy of the skier at the top of the course.

  b) Calculate the kinetic energy of the skier at the bottom of the course.

  c) How would you account for the difference in energy between the two?

10 A 1.5 kg bob on a pendulum is pulled back and raised a height of 0.2 m above its resting position as shown in Figure 2.22.

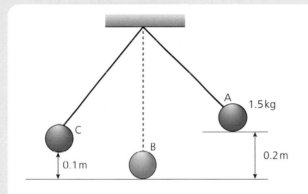

Figure 2.22

a) Calculate its gain in potential energy.

b) Calculate its velocity at B.

c) Calculate E_p and E_k for the bob at C.

11 An 80 kg skier travels down a slope. The starting point is 100 m above the finishing point.

  a) Calculate the gain in potential energy in travelling from the bottom of the slope to the top of the slope.

  b) The skier is travelling at a velocity of $15\,\text{m s}^{-1}$ at the bottom of the slope. Calculate the kinetic energy of the skier.

  c) What would the velocity of the skier be if no energy was lost?

12 A grandfather clock is powered by raising a 2 kg mass a height of 1.5 m and allowing it to fall. As the mass lowers, the arms of the clock turn via a series of springs and gears.

  a) How much energy is gained or stored by raising the mass?

  b) The clock runs for 14 days before the mass needs to be raised. How much energy does it use per second?

Consolidation Questions

1 A 12 kg block is sliding down a slope with a constant acceleration of $6.0\,\text{m s}^{-2}$. The slope is at an angle of 42° to the horizontal.

  a) Calculate the component of the weight acting down in the direction of the slope.

  b) Calculate the frictional force between the block and the surface.

2 A football is rolled off a horizontal bench and bounces along the laboratory floor. The height the ball reaches after each bounce is less than the one before.

Describe the energy changes taking place as the ball bounces along the floor.

Figure 2.23

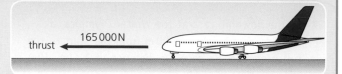

Figure 2.24

3 A student is raising a series of masses by attaching a thin cord to them and raising them to a ledge. The cord is thin and will break if the tension in the cord exceeds 120 N.

He raises a mass of 11.0 kg.

a) What is the greatest upward acceleration he can apply to the mass without the cord breaking?

b) Assuming the mass is raised with the acceleration calculated in part a), what is the shortest time he can take to pull the mass up 7.5 m to the ledge?

4 a) An aircraft of mass 65 000 kg makes its take-off run. The engines produce a thrust of 165 000 N horizontally. Throughout the take-off, frictional forces can be taken as being constant at 35 000 N.

i) Calculate the acceleration of the aircraft along the horizontal runway.

ii) Find the length of the take-off run if the aircraft becomes airborne at a speed of 62 m s^{-1}.

iii) In practice, frictional forces during the take-off run are not constant. Give one reason, with a brief explanation, why this is the case.

b) Shortly after leaving the ground, the aircraft has a nose-up attitude of 15° to the runway. Figure 2.25 shows the forces acting on the aircraft at this time.

i) The aircraft leaves the runway and climbs at an angle of 15° to the runway. Calculate the vertical component of the thrust of the engine during this phase.

ii) Why does the lift force not have a component along the line of flight?

iii) The aircraft travels along its line of flight at a constant speed. Determine the size of the total frictional force opposing its motion now.

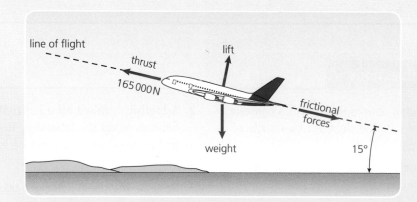

Figure 2.25

5 a) A box of mass 22 kg is at rest on a horizontal frictionless surface.

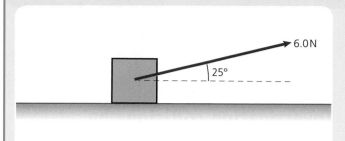

Figure 2.26

A force of 6.0 N is applied to the box at an angle of 25° to the horizontal.

i) Show that the horizontal component of this force is 5.4 N.

ii) Calculate the acceleration of the box along the horizontal surface.

iii) Calculate the time required for the box to be pulled 12.0 m.

b) The box is replaced at rest at its starting position.

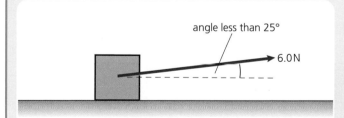

Figure 2.27

The force of 6.0 N is now applied to the box at an angle of less than 25° to the horizontal.

How does the time required by the box to travel 12.0 m compare with your answer to part **a)iii)**? You must justify your answer.

6 An athlete undergoes a training session where his position and time are recorded by a GPS device in his shirt. The athlete sprints 65 m due North then 100 m on a bearing of 135°. The athlete takes 8.5 s to run the first 65 m then another 12.5 s to run the further 100 m.

a) Calculate the displacement of the athlete from the starting point.

b) Calculate the average velocity and average speed of the athlete during this run.

c) The athlete completes another run where he returns to his starting point in another 9.1 s. What is the athlete's average velocity?

3 Collisions, momentum and energy

In general terms, a **collision** refers to a situation where two or more objects come into contact with each other. One can be stationary and the other moving or both can be moving.

We see examples of this phenomenon every day – sports like football, badminton and rugby all have collisions: foot with ball, racket with shuttlecock and defender with forward are all examples of collisions. We also associate collisions with moving vehicles like cars and trolleys but we can expand the principles to include **explosions**. Explosions are where two objects separate by exerting a force on each other. A rocket engine firing out exhaust gases could be considered an explosion. A cannon firing a cannonball would be another example.

To investigate the principles regarding collisions we undertake a series of experiments where small trolleys collide with each other. We measure the masses and velocities of the trolleys before and after a collision and establish any relationships which may be evident.

Investigating collisions by experiment

Example 1

A trolley moving as shown in Figure 3.1 collides with and joins a stationary object.

Immediately after collision the objects (now together) move off at a velocity of $3.75\,\mathrm{m\,s^{-1}}$.

Example 2

Two objects collide as shown in Figure 3.2.

Immediately after collision the objects join together and move off at a velocity of $9.0\,\mathrm{m\,s^{-1}}$.

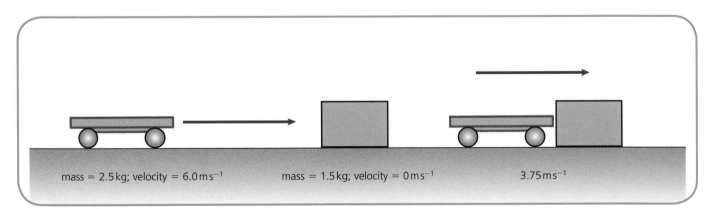

mass = 2.5 kg; velocity = 6.0 m s^{-1} mass = 1.5 kg; velocity = 0 m s^{-1} 3.75 m s^{-1}

Figure 3.1 A collision between a moving object and a stationary object

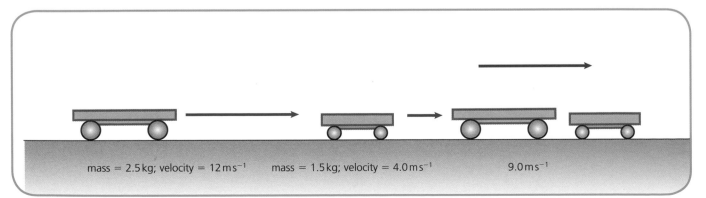

mass = 2.5 kg; velocity = 12 m s^{-1} mass = 1.5 kg; velocity = 4.0 m s^{-1} 9.0 m s^{-1}

Figure 3.2 A collision between two objects moving at different velocities

Is there any relationship that can be drawn from the values of the velocities and the masses?

Calculate the term 'mass × velocity' for each individual object before and after the collision.

Example 1:

Before the collision: $2.5 \times 6.0 = 15$; $1.5 \times 0 = 0$

Total $= 15$

After the collision: 4.0 (as they have combined) $\times 3.75 = 15$

Example 2:

Before the collision: $2.5 \times 12 = 30$; $1.5 \times 4.0 = 6$

Total $= 36$

After the collision: $4.0 \times 9.0 = 36$

The quantity 'mass × velocity' has a numerical value which is the same before a collision and after a collision.

The term 'mass × velocity' is referred to as **momentum**. As velocity is a vector quantity, momentum is therefore a vector quantity. The total momentum before a collision is the same as the total momentum after the collision in the *absence of an external force*. This is often referred to as **conservation of momentum**. It is similar to energy in that it is a concept that we use to describe what we measure and observe. It does not describe fully what happens but allows us to be more detailed in our explanations.

Mass × velocity has units of kg $\times$ m s^{-1} or kg m s^{-1}. There is, however, no specific unit.

The term momentum is used in everyday conversation to try to explain situations. In example 1 the momentum before and after the collision is the same but in conversation people will use phrases like 'the momentum kept the first object moving' or 'the momentum carried through to the second object'. They are not precise enough for an explanation in physics terms but they convey the meaning to some extent. This is normal when specific terms are used in a general way. A physics question where you are asked to explain an event using conservation of momentum will require the answer to be more specific.

In the examples above, the objects continued in the direction they were travelling before the collision. This is not always the case. Sometimes objects stop when they hit another object or rebound in the other direction. The direction is important and the subject of direction leads us back to vectors. Momentum is a vector quantity and direction is crucial to that. If we consider examples where objects are travelling in opposite directions prior to collisions then the sign (positive or negative) of the velocity is important. It does not matter which direction is chosen as the positive one but it has to be consistently applied.

Example 3

Consider the example given in Figure 3.3 in which the two objects colliding are travelling in different directions on a frictionless surface.

The objects stick together after the collision. What is their combined velocity after the collision?

Momentum before:

Take → direction as positive.

Momentum of block (a): $5.0 \times 4.0 = 20 \text{ kg m s}^{-1} \rightarrow$
Momentum of block (b): $2.0 \times -6.0 = -12 \text{ kg m s}^{-1}$

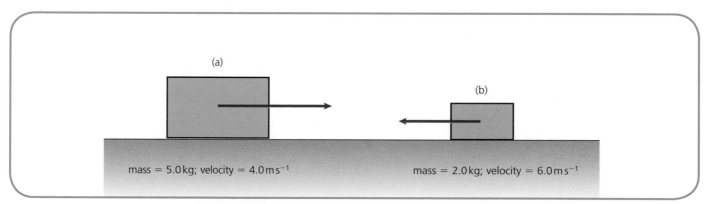

(a)

(b)

mass $= 5.0$ kg; velocity $= 4.0$ m s^{-1}

mass $= 2.0$ kg; velocity $= 6.0$ m s^{-1}

Figure 3.3 A collision between two objects moving in different directions

Total momentum is $20 - 12 = 8.0\,\text{kg}\,\text{m}\,\text{s}^{-1}\rightarrow$ therefore the velocity of the combined block after the collision is calculated using

momentum = mass × velocity	$m \times v$
$8 = 7 \times$ velocity	$7 \times v$
velocity $= 1.14 = 1.1\,\text{m}\,\text{s}^{-1}$	$v = 1.14$

Example 4

An object strikes another object coming in the opposite direction. The masses and velocities are given in Figures 3.4 and 3.5.

Momentum of (a) before: $3.0 \times 4.0 = 12\,\text{kg}\,\text{m}\,\text{s}^{-1}$

Momentum of (b) before: $8.0 \times -2.0 = -16\,\text{kg}\,\text{m}\,\text{s}^{-1}$

Total before $= -4.0\,\text{kg}\,\text{m}\,\text{s}^{-1}$

Momentum after $= -4.0\,\text{kg}\,\text{m}\,\text{s}^{-1}$; momentum of block (a) $= -6.0\,\text{kg}\,\text{m}\,\text{s}^{-1}$. Therefore momentum of block (b) must be $2.0\,\text{kg}\,\text{m}\,\text{s}^{-1}$ (in order to combine to give $-4.0\,\text{kg}\,\text{m}\,\text{s}^{-1}$).

As momentum of $2.0\,\text{kg}\,\text{m}\,\text{s}^{-1}$ = mass × velocity,

velocity of block (b) is $2/8 = 0.25\,\text{m}\,\text{s}^{-1}$ (to the right).

We are only considering here what happens to objects that collide in a straight line but the same principles apply when real-life objects collide: a golfer hitting a golf ball, cars in traffic accidents, a goalkeeper pushing a ball away. The calculations become more complex because of the objects colliding in three dimensions and perhaps travelling quickly but the conservation of momentum applies throughout.

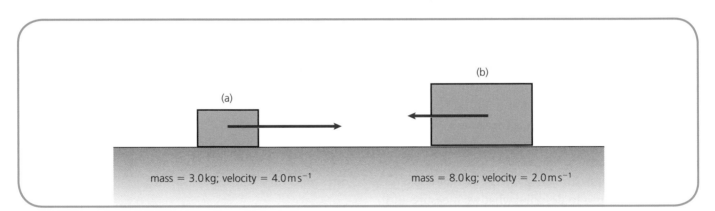

mass = 3.0 kg; velocity = 4.0 ms⁻¹

mass = 8.0 kg; velocity = 2.0 ms⁻¹

Figure 3.4 Before the collision

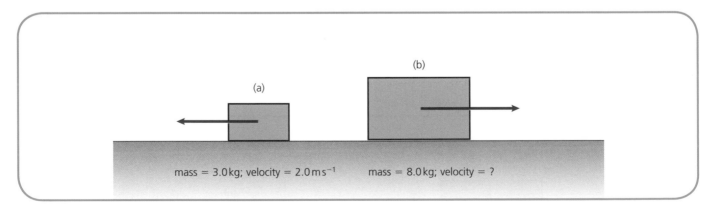

mass = 3.0 kg; velocity = 2.0 ms⁻¹

mass = 8.0 kg; velocity = ?

Figure 3.5 After the collision

Questions

1 A 5.0 kg object travelling at 4.0 m s^{-1} collides with a stationary object of mass 2.0 kg. They stick together and move off.

   a) Calculate the momentum of both objects before the collision.

   b) State the momentum of the objects after the collision.

   c) Calculate the velocity of the objects after the collision.

2 A series of physics experiments involving colliding objects were undertaken by a class. Figure 3.6 shows the masses and velocities of the objects prior to them colliding. In all examples the objects stick together after the collision.

   Calculate their combined velocities after impact.

3 Two ice skaters are moving across the ice at 5.0 m s^{-1}. The man has a mass of 60 kg and the woman has a mass of 80 kg. They push each other apart (in the direction they are moving) and the man's velocity reduces to 3.0 m s^{-1}.

   a) Calculate the momentum they have before separating.

   b) What is the woman's velocity now?

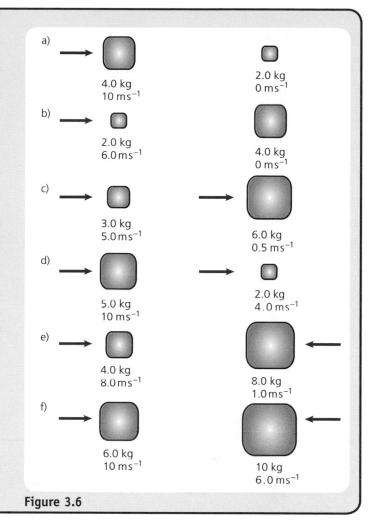

a)
4.0 kg
10 ms^{-1} 2.0 kg
                     0 ms^{-1}

b)
2.0 kg
6.0 ms^{-1} 4.0 kg
                     0 ms^{-1}

c)
3.0 kg
5.0 ms^{-1} 6.0 kg
                     0.5 ms^{-1}

d)
5.0 kg
10 ms^{-1} 2.0 kg
                     4.0 ms^{-1}

e)
4.0 kg
8.0 ms^{-1} 8.0 kg
                     1.0 ms^{-1}

f)
6.0 kg
10 ms^{-1} 10 kg
                     6.0 ms^{-1}

Figure 3.6

Collisions and kinetic energy

In calculating the velocity and momentum associated with collisions so far, discussion of what 'happens' to the energy or where the energy has been transformed has not been mentioned.

We need to consider the energy involved and transformed in these collisions as the objects will have kinetic energy at various points and some energy is transformed when objects are deformed or dented.

Example 1

Consider the following example illustrated in Figure 3.7.

Immediately after the collision the objects (now together) move off at a velocity of 3.75 m s^{-1}.

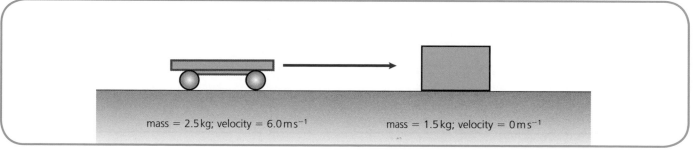

mass = 2.5 kg; velocity = 6.0 ms^{-1} mass = 1.5 kg; velocity = 0 ms^{-1}

Figure 3.7

The objects have kinetic energy before and after the collision.

Kinetic energy before:

$E_k = \frac{1}{2} \times m \times v^2 = \frac{1}{2} \times 2.5 \times 6.0^2 = 45\,J$

$E_k = \frac{1}{2} \times m \times v^2 = \frac{1}{2} \times 1.5 \times 0^2 = 0\,J$

Total = 45 J

Kinetic energy after:

$E_k = \frac{1}{2} \times m \times v^2 = \frac{1}{2} \times 4.0 \times 3.75^2 = 28.1 = 28\,J$

There is less kinetic energy after the collision than there was before so some has been transformed. Some textbooks refer to this as 'lost'. Technically it has not been lost as we cannot lose the energy. It has been transformed into sound or heat or used to deform the objects during the collision.

Example 2

Look at the collision in Figure 3.9 on the next page.

Kinetic energy before:

E_k of (a) $= \frac{1}{2} \times m \times v^2 = \frac{1}{2} \times 5.0 \times 4.0^2 = 40\,J$

E_k of (b) $= \frac{1}{2} \times m \times v^2 = \frac{1}{2} \times 2.0 \times (-6.0)^2 = 36\,J$

Total = 76 J

Kinetic energy after:

$E_k = \frac{1}{2} \times m \times v^2 = \frac{1}{2} \times 7.0 \times 1.1^2 = 4.235\,J = 4.2\,J$

Kinetic energy is not conserved.

In these collisions kinetic energy is not conserved. Momentum is conserved.

Momentum is conserved in all collisions and the total amount of energy (not only kinetic) before and after a collision is conserved. These are the rules or laws that apply. The kinetic energy in these collisions is not conserved; it has been transformed into heat or sound or used to deform some of the object. Collisions of this sort are referred to as **inelastic collisions**.

Worked Example 3.1

Two objects collide as shown in Figure 3.8.

Immediately after the collision the two objects join together and move off at $9\,m\,s^{-1}$.

Kinetic energy before:

$E_k = \frac{1}{2} \times m \times v^2 = \frac{1}{2} \times 2.5 \times 12^2 = 180\,J$

$E_k = \frac{1}{2} \times m \times v^2 = \frac{1}{2} \times 1.5 \times 4^2 = 12\,J$

Total = 192 J = 190 J

Kinetic energy after:

$E_k = \frac{1}{2} \times m \times v^2 = \frac{1}{2} \times 4 \times 9^2 = 162\,J = 160\,J$

Kinetic energy is not conserved.

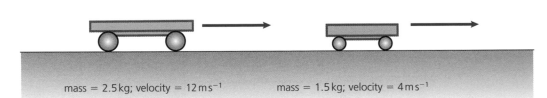

mass = 2.5 kg; velocity = 12 m s⁻¹   mass = 1.5 kg; velocity = 4 m s⁻¹

Figure 3.8

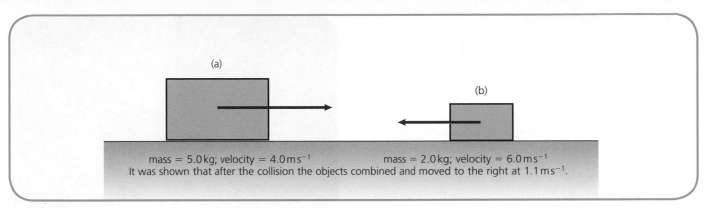

Figure 3.9

Elastic collisions

Consider the example shown in Figure 3.10.

Momentum before $= m \times v = 3.0 \times 5.0 = 15\,\mathrm{kg\,m\,s^{-1}}$;
Momentum before $= m \times v = 3.0 \times 0 = 0\,\mathrm{kg\,m\,s^{-1}}$

Total $= 15\,\mathrm{kg\,m\,s^{-1}}$

To calculate the velocity of the second object after the collision we apply the principle of conservation of momentum.

Momentum after for first object $= 0\,\mathrm{kg\,m\,s^{-1}}$
Momentum after for second object $= 15\,\mathrm{kg\,m\,s^{-1}}$

$m \times v = 3.0 \times v = 15\,\mathrm{kg\,m\,s^{-1}}$

Velocity $= 5.0\,\mathrm{m\,s^{-1}}$

Kinetic energy before collision:

$E_k = \frac{1}{2} \times m \times v^2 = \frac{1}{2} \times 3.0 \times 5.0^2 = 37.5\,\mathrm{J}$
$E_k = \frac{1}{2} \times m \times v^2 = \frac{1}{2} \times 3.0 \times 0^2 = 0\,\mathrm{J}$

Total $= 37.5 = 38\,\mathrm{J}$

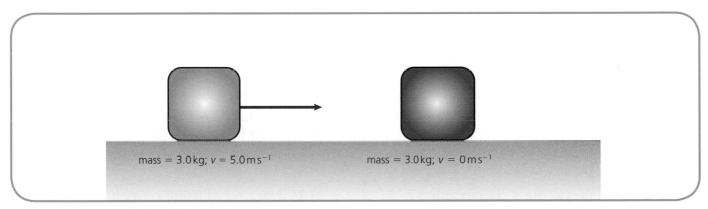

Figure 3.10 Before the collision

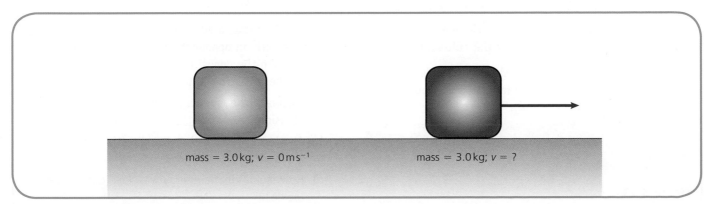

Figure 3.11 After the collision

Kinetic energy after collision:

$$E_k = \frac{1}{2} \times m \times v^2 = \frac{1}{2} \times 3.0 \times 0^2 = 0\,J$$
$$E_k = \frac{1}{2} \times m \times v^2 = \frac{1}{2} \times 3.0 \times 5.0^2 = 37.5\,J$$

Total = 37.5 = 38 J

In this situation the momentum and kinetic energy are both conserved. This is an example of an **elastic collision**.

Elastic collisions hardly ever occur in everyday life. Whenever two objects come into contact there will be energy transformed as heat or sound. Elastic collisions therefore involve objects colliding without touching. Imagine two magnets on a linear air track with similar poles facing each other. The 'repulsion' causes the other object to move but not because it has been touched. When atoms of a gas collide, this approximates closely to an elastic collision as their electrostatic charges can repel other atoms.

Explosive collisions

Explosions can also be described in terms of momentum. While it may seem unusual, an explosion can be thought of as two or more parts of an object separating and pushing off in opposite directions.

Figure 3.12 We can describe such explosions in terms of momentum

Consider two trolleys which are initially at rest and in contact. Their momentum before anything occurs is obviously zero as they are not moving. The kinetic energy of the trolleys is also zero. One trolley releases a spring-loaded piston and both trolleys separate. For our purposes we will only look at objects that separate or explode in one dimension, in other words left/right or up/down. The principle applies in three dimensions but the mathematics becomes a bit cumbersome.

The trolleys separate as shown in Figure 3.14.

Questions

4 In a safety test, a 2.0×10^3 kg car travelling at $6.0\,m\,s^{-1}$ collides with a stationary car of 3000 kg. The cars stick together after the collision.

   a) Calculate the momentum before the collision.

   b) Calculate the kinetic energy before the collision.

   c) Calculate the velocity of the cars after the collision.

   d) Calculate the kinetic energy after the collision.

   e) Explain the difference in kinetic energy calculated before and after the collision.

5 A snooker player attempts a shot. He strikes the cue ball at $2.0\,m\,s^{-1}$. It hits the red ball and stops. They each have a mass of 0.14 kg.

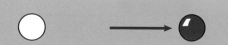

Figure 3.13

   a) Calculate the velocity of the red ball.

A later shot has the cue ball continuing at $0.50\,m\,s^{-1}$ after striking the red ball at $1.6\,m\,s^{-1}$.

   b) Calculate the kinetic energy of the balls before and after the collision.

   c) In reality, not all kinetic energy is conserved and transferred to the red ball. What has this energy been transformed into?

6 In a rugby tackle a 90 kg player running at $5\,m\,s^{-1}$ collides with an opponent of 110 kg running at $4.0\,m\,s^{-1}$ in the opposite direction. They hold on to each other after the collision.

   a) Calculate the momentum of the players before the collision.

   b) At what speed and in what direction do they move after the collision?

   c) How much kinetic energy was 'lost' during the tackle?

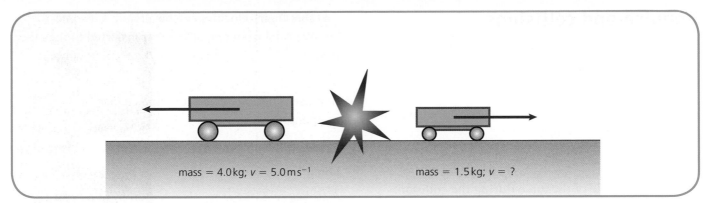

Figure 3.14

Momentum before the explosion is zero. If we apply the principle of conservation of momentum then the momentum after the explosion is also zero. This is consistent as momentum is a vector quantity and in effect the momentum in one direction is balanced or cancelled out by the momentum in the other direction.

Momentum to left $= m \times v = 4.0 \times (-5.0)$
$$= (-20)\,\text{kg}\,\text{m}\,\text{s}^{-1}$$

Total momentum equals zero so momentum to right must be $20\,\text{kg}\,\text{m}\,\text{s}^{-1}$.

$m \times v = 20\,\text{kg}\,\text{m}\,\text{s}^{-1}$; $v = 20/1.5 = 13.33 = 13\,\text{m}\,\text{s}^{-1}$

Kinetic energy before:

$E_k = 0$ as objects are stationary.

Kinetic energy after:

$E_k = \frac{1}{2} \times m \times v^2 = \frac{1}{2} \times 4.0 \times (-5.0)^2 = 50\,\text{J}$

$E_k = \frac{1}{2} \times m \times v^2 = \frac{1}{2} \times 1.5 \times 13.3^2 = 132.7 = 130\,\text{J}$

Total $= 180\,\text{J}$

This is not an elastic collision as the kinetic energy prior to the explosion is zero. The kinetic energy is due to the transformation of the energy 'stored' in the spring or piston as the trolleys were put together.

These 'explosions' embody the principle of conservation of momentum and this principle applies to many circumstances. A rocket's combustion chamber generates a huge mass of material which is ejected downwards and this causes the rocket to be projected upwards. Similar principles apply to shells or bullets being fired or a jet engine ejecting a large mass of air at a great velocity in order to push a large mass of aluminium and carbon composite (and people) in the opposite direction.

Figure 3.15 A bullet firing illustrates the principle of conservation of momentum

Worked Example 3.2

An electrical power worker needs to fire a length of cable across a small river. A cable is attached to the projectile and it is fired across the river. The cable and casing have a mass of 1.2 kg and leave the launcher at $45\,\text{m}\,\text{s}^{-1}$.

The worker and launcher have a combined mass of 95 kg. Calculate the velocity at which they recoil.

Momentum before = zero

Momentum after = zero

Momentum of cable/casing = mass × velocity
$= 1.2 \times 45 = 54\,\text{kg}\,\text{m}\,\text{s}^{-1}$

Therefore momentum of worker and launcher must equal $-54\,\text{kg}\,\text{m}\,\text{s}^{-1}$. This means $95 \times v = -54$. This leads to $v = -54/95 = -0.57\,\text{m}\,\text{s}^{-1}$.

The worker and launcher recoil at $0.57\,\text{m}\,\text{s}^{-1}$.

Impulse and collisions

An area that has not been considered so far is what happens when an object is acted upon by a force (something has been struck!).

To appreciate what happens in a collision we have to consider two main factors.

● The duration of the collision.

● The magnitude of the force during the collision.

These factors are also identified if we consider a mathematical analysis of the issues.

Take the equations:

$$F = m \times a$$

$$a = \frac{(v - u)}{t} \text{ (from } v = u + at)$$

Combining the two gives $F = m\dfrac{(v - u)}{t}$

Which leads to $F \times t = m(v - u)$.

This can be thought of as the following: 'The force applied to an object for a (multiplied) certain time is equivalent to the mass multiplied by the change in velocity.'

This leads to the idea that 'The force multiplied by the time it acts = mass multiplied by change in velocity.'

This can be described as

$$F \times \Delta t = m \times \Delta v$$

When a collision occurs, the effect of that collision can be estimated by multiplying the force by the time it acts for and this is equivalent to the mass × change in velocity.

The force multiplied by the time it acts is therefore equal to the change in momentum of the object.

This concept is referred to as **impulse** and it can be considered that in a collision, the impulse from one object is transferred to the other. This can also be explained using our 'conservation of momentum' principle. Consider two objects about to collide. We know that the momentum before the collision is the same as the momentum after the collision, yet the two objects can change their velocities. As the overall momentum of the system does not change, the only

way that their velocities can be altered is if some 'momentum' from one object is transferred to the other. In trying to explain this we use the term or concept impulse.

Impulse is often referred to as the change in momentum.

Examples to illustrate the principle of impulse

Example 1

A trolley of 4.0 kg at rest is acted on by a force of 15 N for 3.0 seconds.

a) Calculate the change in momentum of the trolley.

$$F \times \Delta t = m \times \Delta v$$

$$15 \times 3.0 = \text{impulse (change in momentum)} = 45 \, \text{N s}$$

b) Calculate its final velocity.

$$45 \, \text{N s} = m \times \Delta v$$

$$45 \, \text{N s} = 4.0 \times \Delta v$$

$$\Delta v = 11.25 \, \text{m s}^{-1}$$

Final velocity $= 11.25 = 11 \, \text{m s}^{-1}$

Example 2

A golf ball is struck by a driver. The head of the driver has a mass of 0.50 kg. The ball has a mass of 0.045 kg. The club head strikes the ball and is in contact with the ball for 0.80 ms. It applies an average force of 2500 N while it is in contact with the ball.

At what velocity does the ball leave the club?

$$F \times \Delta t = m \times \Delta v$$

$$2500 \times 0.0008 = 0.045 \times \Delta v$$

$$\Delta v = 44 \, \text{m s}^{-1}$$

Example 3

A car of mass 1800 kg is travelling at 15 m s^{-1}. In a test it must stop within 12 seconds. What force must the brakes apply in order for this to happen?

In this case the change in velocity is $-15 \, \text{m s}^{-1}$ as the car must stop.

$$F \times \Delta t = m \times \Delta v$$

$$F \times 12 = 1800 \times (0 - 15)$$

$$F = -2250 = -2300 \, \text{N}$$

Therefore, the brakes must apply a force of –2300 N in order to stop in this time.

If we want to create a great impact or a greater transfer of momentum we need to have a large force and it must be in contact with the object for a long time. An example of this could be bobsleighing sprinters as they start their run. They push their hardest for as long as possible until they have to get into the vehicle. This application of a large force for as long as possible ensures there is a large change in momentum of the bob.

Most collisions tend to be of a large force but in contact for a short time. Sporting collisions of balls or shuttlecocks, for example, involve objects being in contact for a short time but applying a relatively large force when in contact.

In reality the force applied to an object is never the same throughout the period of contact. As an object

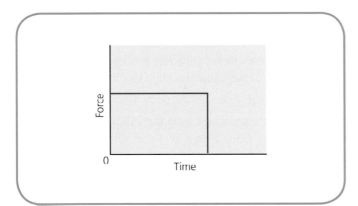

Figure 3.16 An ideal force–time graph

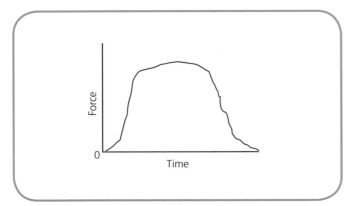

Figure 3.17 When we consider the graphs, the term $F \times \Delta t$ is the area under the force–time graph. This is equivalent to the impulse. We can solve more complex questions if we can calculate this area

makes contact, the initial 'touch' is a small force which increases to a larger or maximum force and then decreases to zero as the objects separate. If we look at the force–time graph for our collisions, an ideal one would look like the graph shown in Figure 3.16 and a 'real' one could look like Figure 3.17.

A hammer strikes a nail of mass 35 g and pushes it into a block of wood. The head of the hammer has a mass of 400 g and makes contact with the nail. The force–time graph is given as Figure 3.18.

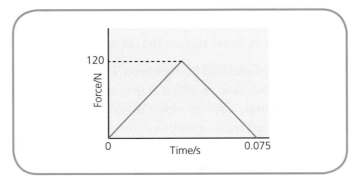

Figure 3.18

Calculate the impulse passed to the nail.

Impulse = $F \times \Delta t$ = area under F–t graph

Area = $\frac{1}{2} \times 0.075 \times 120 = 4.5\,\text{N s}$

Calculate the velocity of the nail immediately after impact.

$F \times \Delta t = m \times \Delta v$

$4.5 = 0.035 \times v$

$v = 4.5/0.035 = 128 = 130\,\text{m s}^{-1}$

This is obviously a bit on the quick side, but as the nail enters the wood it forces aside the particles of the wood, compressing them to the surrounding wood. The compressed wood is difficult to push through and this friction 'grips' the nail in the wood very strongly. This friction also slows down the nail very quickly, in the space of a few cm generally.

Pile drivers operate on the same principle. A large mass is raised in a long steel column and then dropped from a height. It falls down the column and collides with a large steel or concrete pile which is gradually hammered into the ground. The friction between the pile and the earth is great and it is very difficult to hammer the pile in. It is this friction that supports

49

the weight of the structure. If we were just to build on top of the soil, it would compress and one end of the building could sink lower than the other end. This would cause lots of cracks in walls and can be unsafe. Many old buildings exhibit these cracks.

When objects interact they exert a force on each other. Many people refer to this as 'For every action there is an equal and opposite reaction.' However, this is inaccurate as, for example, closing one door does not make another one open. The two forces need to be interacting. An object placed on a table will exert a downwards force on the table and the table exerts an upwards force of equal size on the object.

This is best summarised by **Newton's Third Law**: '*If object A exerts a force on object B then object B exerts an equal and opposite force on object A*'. Note that this is the version expected as a response in assessments.

This idea of mutual, equal and opposite forces may be stated mathematically as:

$$F_1 = -F_2$$

The impulse equation then gives

$$\Delta(m_1v_1)/\Delta t = -\Delta(m_2v_2)/\Delta t$$

Cancelling out time of contact

$$\Delta(m_1v_1) = -\Delta(m_2v_2)$$

Rearranging gives

$$\Delta(m_1v_1) + \Delta(m_2v_2) = 0$$

This is a statement of the law of conservation of momentum, i.e. in the absence of external forces the total momentum before and after an interaction (collision or explosion) is conserved. This conservation law is directly derived from Newton's Third Law.

Worked Example 3.3

In manned space missions to the International Space Station (ISS), supply craft and space shuttles had to dock with the ISS in order for the crew and goods to transfer from one to the other.

Figure 3.19

The movement of the supply craft and shuttles is controlled by small rocket thrusters positioned throughout the spacecraft. These allow the craft to rotate, move away and so on.

When the shuttle separated from the ISS in order to return, it had to move a mass of 1.0×10^5 kg safely away from the ISS.

When leaving, the shuttle's rocket thrusters fire and a constant force of 2750 N is applied to the shuttle.

How long must the thrusters act in order for the shuttle to reach a velocity of 2.0 m s^{-1}?

$F \times \Delta t = m \times \Delta v$

$2750 \times \Delta t = 1.0 \times 10^5 \times 2.0$

$\Delta t = 1.0 \times 10^5 \times 2.0/2750 = 73$ seconds

Questions

7 A 4 kg shell explodes into two sections. A 1.5 kg section moves to the right at 28 m s^{-1}. Calculate the velocity of the other section.

8 A 500 kg mass is raised and dropped onto metal piles to drive them into the ground. The mass is raised 3 m above the pile and released. The pile

→

has a mass of 3000 kg. The mass and the pile stick together after impact.

a) Calculate the downwards velocity of the mass as it strikes the pile.

b) Calculate the initial downwards velocity of the pile and mass after impact.

c) Calculate the kinetic energy of the pile and mass just after impact.

9 In a circus trick a man of 80 kg is fired from a cannon into a net some distance away. The performer leaves the cannon at a velocity of 15 m s⁻¹ to the right. The cannon has a mass of 500 kg.

a) Calculate the velocity at which the cannon recoils.

b) The piston in the cannon is in contact with the 'human cannonball' for 0.5 seconds as it fires him out. Calculate the average force experienced by him.

10 A golf ball is struck and leaves the club head at 50 m s⁻¹. The mass of the ball is 0.045 kg.

a) Calculate the momentum of the golf ball after the strike.

b) If the ball was in contact with the club head for 5 ms, calculate the average force applied to the ball.

c) How does the speed of the golf club change during the stroke?

Consolidation Questions

1 a) A shell is fired from a field gun and leaves the gun with a velocity of 160 m s⁻¹. The barrel of the gun is inclined at an angle of 20° to the horizontal as shown in the diagram.

Figure 3.20

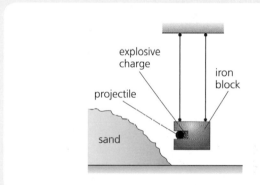

Figure 3.21

Determine the horizontal and vertical components of the shell's initial velocity.

b) The procedure described below is used to test explosives for propelling artillery shells.

A large iron block is suspended as a pendulum (Figure 3.21). The explosive charge is packed into a hole in the side of the iron block and a suitable projectile put in after the explosive. When the explosive is detonated, the projectile is fired out and lands harmlessly in the sand while the pendulum swings back and the maximum height it reaches is measured as shown in Figure 3.22.

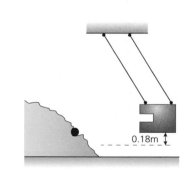

Figure 3.22

In one such test, the following results were recorded:

Mass of pendulum = 275 kg

Mass of projectile = 7.5 kg

Height risen by pendulum = 0.18 m

i) Find the horizontal velocity of the pendulum immediately after the explosion.

ii) Hence find the velocity of the projectile as it left the pendulum.

iii) Find the average force exerted on the projectile by the explosive charge if the explosion lasted for 20 ms.

iv) Devise and describe a safe method of measuring the vertical distance moved by the pendulum.

2 A student uses a linear air track to investigate collisions. In one experiment a vehicle of mass 0.55 kg moves along and rebounds from a metal spring mounted at one end of the level track as shown below.

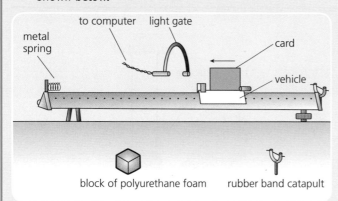

Figure 3.23

By using a light gate connected to a computer, she obtains values for the speed of the vehicle before and after it collides with the spring.

She then repeats this procedure, replacing the metal spring first with the block of polyurethane foam and then with the rubber band catapult. She records the results of each experiment in a table as shown in Table 3.1.

a) Calculate values of kinetic energy to complete the last row of the table.

b) For which experiment is the collision most nearly elastic? You must justify your answer.

c) Describe a method she could use to give the vehicle the same initial speed each time.

3 **a)** A bullet of mass 32 g is fired horizontally into a sand-filled box which is suspended by long strings from the ceiling. The combined mass of the bullet, box and sand is 7.5 kg.

After impact, the box swings upwards to reach a maximum height as shown in Figure 3.24.

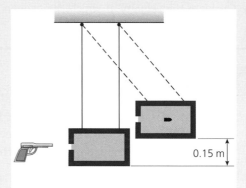

Figure 3.24

Calculate:

i) the maximum velocity of the box after impact

ii) the velocity of the bullet just before impact.

b) The experiment is repeated with a metal plate fixed to one end of the box as shown in Figure 3.25.

The mass of sand is reduced so that the combined mass of the sand, box and metal plate is 10 kg.

In this experiment, the bullet bounces back from the metal plate. Explain how this would affect the maximum height reached by the box compared with the maximum height reached in part **a)**.

	Metal spring	Polyurethane block	Rubber band
Speed before collision/m s⁻¹	0.62	0.62	0.62
Speed after collision/m s⁻¹	0.58	0.41	0.47
Kinetic energy before collision/J	0.106	0.106	0.106
Kinetic energy after collision/J	0.092		

Table 3.1

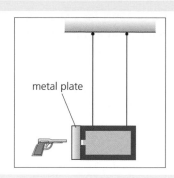

Figure 3.25

4* The apparatus shown below is used to test concrete pipes.

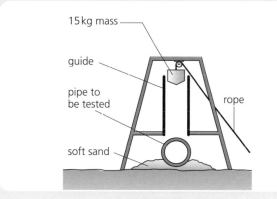

Figure 3.26

When the rope is released, the 15 kg mass is dropped and falls freely through a distance of 2.0 m onto the pipe.

a) In one test, the mass is dropped onto an uncovered pipe.

   i) Calculate the speed of the mass just before it hits the pipe.

   ii) When the 15 kg mass hits the pipe, the mass is brought to rest in a time of 0.02 s. Calculate the size and direction of the average unbalanced force on the pipe.

b) The same 15 kg mass is now dropped through the same distance onto an identical pipe which is covered with a thick layer of soft material. Describe and explain the effect this layer has on the size of the average unbalanced force on the pipe.

5 A space vehicle of mass 2750 kg is moving with a constant speed of 0.55 m s^{-1} in the direction shown in Figure 3.27. It is about to dock with a space probe of mass 1250 kg which is moving with a constant speed in the opposite direction.

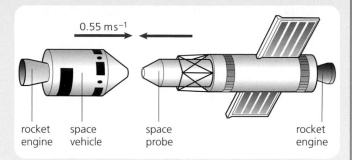

Figure 3.27

After docking, the space vehicle and space probe move off together at 0.25 m s^{-1} in the original direction in which the space vehicle was moving.

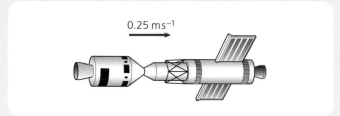

Figure 3.28

a) Calculate the speed of the space probe before it docked with the space vehicle.

b) The space vehicle has a rocket engine which produces a constant thrust of 1200 N. The space probe has a rocket engine which produces a constant thrust of 650 N. The space vehicle and space probe are now brought to rest from their combined speed of 0.25 m s^{-1}.

   i) Which rocket engine was switched on to bring the vehicle and probe to rest?

   ii) Calculate the time for which this rocket engine was switched on. You may assume that a negligible mass of fuel was used during this time.

c) The space vehicle and space probe are to be moved from their stationary position at A and brought to rest at position B, as shown.

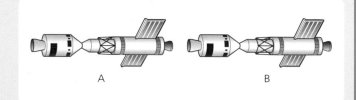

Figure 3.29

Explain clearly how the rocket engines of the space vehicle and the space probe are used to complete this manoeuvre. Your explanation must include an indication of the relative time for which each rocket engine must be fired. You may assume that a negligible mass of fuel is used during this manoeuvre.

6 Two ice skaters are initially skating together, each with a velocity of $2.5\,\text{m}\,\text{s}^{-1}$ to the right as shown in Figure 3.30.

The mass of skater R is 55 kg. The mass of skater S is 45 kg. Skater R now pushes skater S with an

average force of 120 N for a short time. This force is in the same direction as their original velocity. As a result, the velocity of skater S increases to $4.5\,\text{m}\,\text{s}^{-1}$ to the right.

a) Calculate the magnitude of the change in momentum of skater S.

b) How long does skater R exert the force on skater S?

c) Calculate the velocity of skater R immediately after pushing skater S.

d) Is this interaction between the skaters elastic? You must justify your answer by calculation.

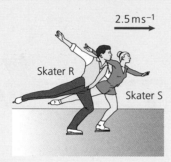

$2.5\,\text{ms}^{-1}$

Skater R

Skater S

Figure 3.30

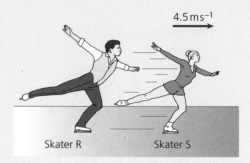

$4.5\,\text{ms}^{-1}$

Skater R

Skater S

Figure 3.31

4 Projectiles and satellites

Projectiles

A **projectile** is the term we give to an object that has been launched, thrown or released into the air.

Once it leaves our launcher or hand it will move away and fall to Earth. While in the air the forces acting on it are gravity and any air friction or drag. This differs from what was discussed earlier in that we now consider what happens in two dimensions.

When a projectile is released it moves away from its initial position and travels along in a horizontal direction and also travels in a vertical direction due to the influence of gravity. This combination of a horizontal and vertical motion leads to the standard shape we are familiar with when we observe balls or stones, for example, flying through the air. This curved path is shown in Figure 4.1.

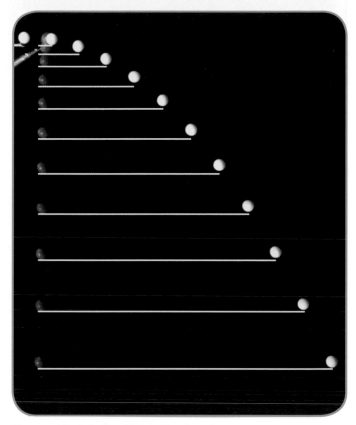

Figure 4.2 A ball rolling off a bench

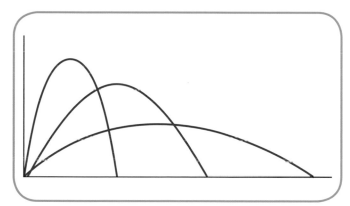

Figure 4.1 The familiar motion of objects being launched from a point

If we look closely at the motion of the projectiles, we can discern some common themes.

Close inspection of the photograph in Figure 4.2 shows that the distance travelled in the horizontal direction is the same after every flash and that the horizontal velocity of the object has no real impact on its vertical velocity! This seems counterintuitive; it is, however, true. If we ignore air friction, the only factor that determines the time taken for an object to fall is the height from which it is dropped.

This leads to the following:

- A ball that rolls off a bench at a high velocity will take the same time to hit the ground as one that merely tips over.

- A tennis ball that is struck horizontally at a very high velocity will hit the ground in the same time as one that has been dropped from the same height.

- A bullet fired horizontally from a rifle will hit the ground in the same time as one that is dropped from the same height.

If the horizontal velocity has no effect on its vertical acceleration then when calculating how far a ball goes or how long it takes to travel a certain distance, we can separate the horizontal and vertical components of the motion and treat them separately.

Worked Example 4.1

A ball rolls horizontally off a bench of height 1.2 m. It has an initial horizontal velocity of 3.0 m s^{-1}.

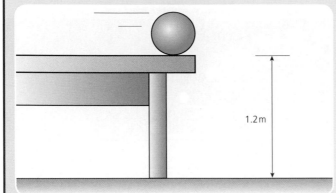

Figure 4.3

How far away from the bench does it land?

The determining factor is the height. Consider it as if the ball falls from a height of 1.2 m. To calculate the time it

takes to fall we use our equations of motion. Considering only the vertical motion:

v = ?, u = 0, a = 9.8, t = ?, s = 1.2

As the ball is falling downwards (and gravity acts in that direction) we do not need to use a sign convention. To calculate t we use the equation

$$s = ut + \frac{1}{2}at^2$$

This leads to $s = \frac{1}{2}at^2$ as u = 0.

This gives $1.2 = \frac{1}{2} \times 9.8t^2$; solving for t gives t = 0.5 s. The ball is in the air for 0.5 s.

This is the time that the ball is in the air for. It has a horizontal velocity of 3.0 m s^{-1}. This translates into a horizontal distance of

$$s = v \times t = 3.0 \times 0.5 = 1.5\,\text{m}$$

The ball hits the floor 1.5 m from the bench.

Worked Example 4.2

A stone is thrown horizontally off a cliff at a velocity of 5.0 m s^{-1}. It splashes into the water 3.5 s later.

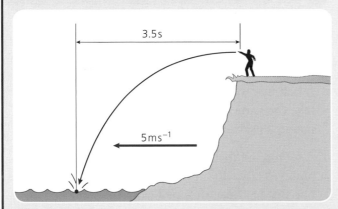

Figure 4.4

a) How far away from the cliff did the stone enter the water?

Horizontal displacement = $v \times t$ = 5.0 × 3.5 = 17.5 m

b) How high above the water was the stone when it was thrown?

The stone took 3.5 s to hit the water. This means it fell under gravity for 3.5 s. Equations of motion can calculate the displacement.

$$s = ut + \frac{1}{2}at^2$$

This leads to $s = \frac{1}{2}at^2$ as u = 0 (vertically)

This gives $s = \frac{1}{2} \times 9.8 \times 3.5^2 = 60\,\text{m}$

c) What is the velocity of the stone as it enters the water?

Its vertical velocity can be calculated using

$$v = u + at = 0 + 9.8 \times 3.5 = 34.3\,\text{m s}^{-1}$$

Its horizontal velocity is 5.0 m s^{-1} so we combine these velocities in the same manner as we combined vectors earlier using Pythagoras' theorem:
$a^2 + b^2 = c^2$, so $5.0^2 + 34.3^2 = c^2$

c = 34.7 = 35 m s^{-1}

The angle at which the stone enters the water can be calculated using simple trigonometry.

$\tan\theta = 5.0/34.7 = 0.14$

θ = 8.2°

The stone enters the water at 35 m s^{-1} at an angle of 8.2° from vertical.

Worked Example 4.3

A ball is struck and leaves the ground as shown.

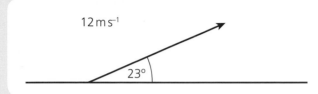

Figure 4.5

a) Calculate the horizontal and vertical components of its velocity.

$$H = 12 \times \cos 23° = 11\,\mathrm{m\,s^{-1}}$$
$$V = 12 \times \sin 23° = 4.7\,\mathrm{m\,s^{-1}}$$

b) Calculate the time the ball is in the air until its first bounce.

This is calculated using the vertical component of the velocity. When the ball returns to the ground its vertical displacement will be 0 m. Its vertical velocity will be of the same magnitude but in the opposite direction.

$$v = -4.7\,\mathrm{m\,s^{-1}},\ u = 4.7\,\mathrm{m\,s^{-1}},\ a = -9.8\,\mathrm{m\,s^{-2}},$$
$$t = ?,\ s = 0\,\mathrm{m}$$
$$a = (v - u)/t$$
$$-9.8 = (-4.7 - 4.7)/t$$
$$-9.8 = -9.4/t$$
$$t = -9.4 / -9.8 = 0.96\,\mathrm{s}$$

c) Calculate the magnitude of the horizontal displacement.

This is calculated by multiplying the horizontal component of the velocity by the time the ball is in the air until its first bounce.

$$s = \text{horizontal velocity} \times t$$
$$= 11 \times 0.96$$
$$= 10.56 = 11\,\mathrm{m}$$

Questions

1 A stone is dropped off a cliff and lands in the water 4.0 s later.

  a) How high is the cliff?

  b) At what velocity does the stone hit the water?

2 A cylinder rolls off a bench with a horizontal velocity of 2.0 m s⁻¹. The bench is 1.5 m high.

  a) How long does it take to hit the ground?

  b) What is the vertical component of velocity on impact?

  c) How far from the bench does it land?

  d) At what angle does it strike the ground?

3 A cannonball is fired horizontally from a cliff edge. It leaves the cannon with a velocity of 80.0 m s⁻¹. The cliff is 25 m above the water level.

  a) How long does it take for the cannonball to fall 25 m?

  b) How far away from the cliff does the ball hit the water?

  c) What angle does the ball make with the horizontal when it lands?

4 A ball rolls down a slope and into freefall as shown in Figure 4.6.

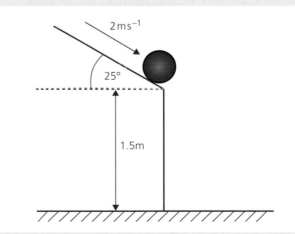

Figure 4.6

  a) Calculate the horizontal and vertical components of the ball's velocity at the moment it leaves the slope.

  b) How long does it take to reach the ground?

  c) How far away does it land from the base of the slope?

Objects moving quickly

Many people assume that an object that moves very quickly is somehow not affected by gravity. For example when a bullet flies horizontally, the moment it leaves the barrel gravity acts upon it and it starts to fall down. But since it is travelling very, very quickly, to our eyes – or even using a slow motion camera – the bullet appears to be travelling horizontally. This is due to the fact that it passes across our field of view, or the distance to the target, so quickly that it may only travel a few millimetres (or less) in the downward direction. This distance is too small for us to notice.

The same may be true of football shots when a player scores from 10 m or so. If the ball is struck firmly it will travel from the player's foot to the roof of the net in a fraction of a second. In reality, the ball is travelling upwards but with a slight curving downwards of perhaps a centimetre or two. As this slight curve downwards is hard to see, we consider the ball to have travelled in a straight line without any gravitational effect. If we look at a goalkeeper who takes a goal kick, the ball is struck very hard and flies off in an apparently straight line but eventually it returns to Earth. Projectiles start to move with a downward component the moment they are free from the force causing their motion. Golf shots or balls thrown all do the same.

These examples show that all projectiles will 'move' downwards under the influence of gravity. Their horizontal velocity will carry them further horizontally before they hit the ground, but they will hit the ground in the same time as if they all started from the same height.

Figure 4.7 shows what happens if an object moves with increasing horizontal velocity. The objects will go further and their curvature will gradually be lessened.

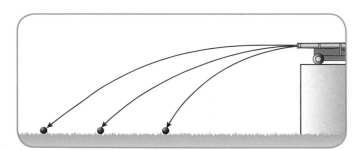

Figure 4.7

Satellites

Consider a situation in which there was no air friction or drag and the object had such a great horizontal velocity that its curve was almost horizontal. Consider also that this object is moving so quickly that as it falls, the curvature of the Earth 'falls' or curves away at the same rate as the object.

As there is no air resistance, the object does not slow down. It keeps moving horizontally and falling but the Earth is curving round as it does so. The object never falls to the ground. It falls completely around the Earth. In other words, it orbits the Earth.

This is why **satellites** and other space vehicles remain around or above the Earth. They must be travelling with an appropriate horizontal velocity which allows them to fall at the same rate as the surface of the Earth falls away or curves.

The orbital velocity of the satellite depends on its altitude above Earth. The closer to the Earth the satellite is, the greater the velocity at which it must travel. At an altitude of 124 miles (200 kilometres) above the Earth's surface, the required orbital velocity is just over 17 000 mph (about 27 400 kph).

To maintain an orbit that is 22 223 miles (35 786 km) above Earth, the satellite must orbit at a velocity of about 7000 mph (11 300 kph). At that particular height above the surface, orbital velocity and corresponding 'circumference' the satellite will make one complete revolution in 24 hours. Since the Earth also rotates once in 24 hours, a satellite at 22 223 miles (35 786 km)

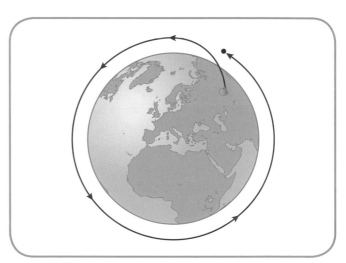

Figure 4.8 Path of a satellite entering orbit

altitude can remain in a fixed position relative to a point on the Earth's surface (above the equator). This kind of orbit is called a **geostationary orbit**. Such orbits are ideal for weather satellites and communications satellites as they remain at a set point relative to the surface and can 'view' or communicate with stations or receivers in that region without the receiver having to continually point to a different area of the sky.

Geostationary satellites generally orbit at some point above the equator. They cannot orbit above Scotland, for example, as they must orbit above the equator of the Earth.

Many people are of the opinion that satellites and other space vehicles just hover or 'float' in space. This is not the case. If the vehicles did not have a horizontal velocity, they would accelerate towards the Earth in a straight line. Satellites that slow down, such as old satellites for example, start to curve towards the Earth more than the Earth curves away. In effect, this means they get closer to the surface with each orbit.

As they fall further towards the Earth, they start to make contact with some air particles in the upper atmosphere which in turn slows them down even more and they make even more contact with the air. Such satellites eventually enter the Earth's atmosphere and evaporate due to the heat generated by the air friction. On occasion, not all of the debris evaporates and some large fragments crash to Earth.

If you can travel at a great enough velocity, the amount you curve down due to gravity is less than the curving

of the Earth. At this point you will appear to gain height from the ground and you eventually leave or escape the Earth's gravitational field. This velocity for Earth is approximately 25 000 mph and is referred to as the **escape velocity** for Earth.

Gravitational field strength

The attraction that objects have for each other caused only by their mass has concerned scientists and philosophers for many years. Various theories and explanations have been put forward to explain what is observed. All can explain the same phenomenon but at various levels of complexity.

What is accepted is that an object which has a mass has an associated gravitational field. This means that other objects which come relatively close to that object will experience a force of attraction and the objects will move towards each other. As the mass of an object increases, the gravitational field associated with it will also increase.

It is generally agreed that the large clouds of dust formed after the Big Bang gradually aggregated together to form larger and larger masses. As the masses increased, their own gravitation exerted huge pressures on the atoms and molecules in the core. As this process continues, and if certain conditions are met, the atoms at the centre will undergo 'fusion' and begin to radiate large amounts of energy. This is the basic mechanism by which stars form.

As stars form, a disc of dust-like matter can also appear and these discs tend to rotate around the centre of the

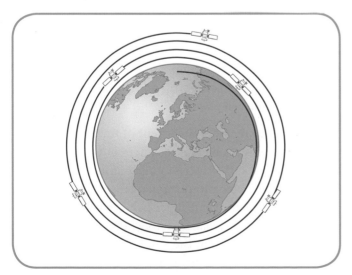

Figure 4.9 Satellites that slow down get closer to the Earth's surface as they orbit

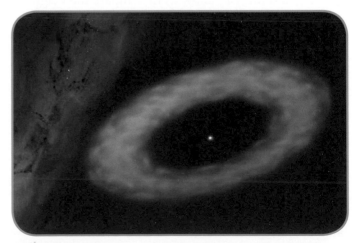

Figure 4.10 An early star with its associated disc, the precursor of a planet

star. Over time this dust forms its own larger masses and this is the mechanism that we believe leads to the formation of planets.

Larger planets have a greater mass and therefore a greater gravitational field. We use the term **gravitational field strength** to compare the force on a mass of 1 kg at various positions on planets.

The gravitational field strength on Earth is 9.81 newtons per kg ($N kg^{-1}$). A 1 kg mass will experience a force of 9.81 N on the surface of the Earth. This force on the mass is its weight. Weight is a force and is measured in newtons. In scientific terms, our own weight should be measured in newtons but it is generally accepted that people use kg or stones. It is incorrect but this habit is difficult to break.

The gravitational field strengths of our Sun and some associated bodies are given in Table 4.1.

Body	Gravitational field strength/$N kg^{-1}$
Sun	274.13
Mercury	3.59
Venus	8.87
Earth	9.81
Moon	1.62
Mars	3.77
Jupiter	25.95
Saturn	11.08

Table 4.1

The weight of a 1 kg mass can be calculated using $W = mg$ and this can allow us to compare the weight of a probe (or astronaut) if we were ever to send one to the surface of these planets.

The force experienced by objects in proximity to larger masses obviously depends upon the mass of each of the objects and also the distance between them. A mass on the surface of a planet will experience a greater force on it than one which is some distance away.

Newton's Law of Universal Gravitation

Sir Isaac Newton proposed that the force of attraction 'caused by gravity' between two objects could be calculated by the formula

$$F = G \frac{m_1 m_2}{r^2}$$

where:

- F is the force between the masses
- G is the **gravitational constant**
- m_1 is the first mass
- m_2 is the second mass
- r is the distance between the masses.

This inverse square relationship was suggested by a number of people prior to Newton's publication, but he brought a level of accuracy and detail to this relationship that had been lacking. He did not, however, suggest what caused this 'action at a distance' or why the force existed.

This relationship for calculating the forces of attraction between objects works well for the most part but it loses a lot of its accuracy when we deal with how light is affected by gravity, for example, and we need to consider relativity to more accurately explain our observations.

The formula $F = G \dfrac{m_1 m_2}{r^2}$ helps to describe the attraction felt by masses due to other masses which can be some distance away. The Moon orbits the Earth due to the attraction that each body experiences. We do not feel this attraction but it is responsible, in combination with the rotation of the Earth, for raising and lowering sea levels as the Moon moves around us. Tides are complex and difficult to fully understand. The Earth's solid material, such as the crust, is also attracted by the gravitational pull of the Moon but its movement is much smaller (around 30 cm or so).

All the objects in our Solar System exert forces on each other and these forces all combine to cause the movement of planets, moons and asteroids. The Sun is obviously the dominant body and its influence extends way beyond the last planet.

It is difficult to measure the extent of the Sun's gravitational effects at great distances but the Kuiper belt, which is about 40 astronomical units from the Sun, may give a better indication of the real magnitude of the Solar System. The Kuiper belt is a region in the distant edges of the Solar System where many objects reside. They are far from the Sun but are kept there due to the gravitational attraction of the Sun. (An astronomical unit is the mean distance from the Sun to the Earth.)

Worked Example 4.4

Mass of Earth: 5.97×10^{24} kg
Mean radius of Earth: 6.37×10^6 m
Mean distance to the Moon: 3.84×10^8 m

The Gravitational constant, G, has a value of 6.67×10^{-11} m^3 kg^{-1} s^{-2}.

Calculate the force of attraction, from the Earth, a 100 kg mass would experience if it was placed at the same distance from the Earth as the Moon.

$F = (Gm_1 \times m_2)/r^2$
$= (6.67 \times 10^{-11} \times 100 \times 5.97 \times 10^{24})/(3.84 \times 10^8)^2$
$= 3.98 \times 10^{16}/1.47 \times 10^{17}$
$= 0.27$ N

The force of attraction between this mass and the Earth is 0.27 N. This would result in an acceleration of $F/m = 0.0027$ m s^{-2} towards the Earth.

Worked Example 4.5

Calculate the force of attraction of the Earth on a 70 kg astronaut currently working in the space station. The space station orbits at a height of 430 km above the Earth.

The distance from the astronaut to the Earth is taken from the centre of the Earth.

$r = 6.37 \times 10^6 + 4.3 \times 10^5 = 6.8 \times 10^6$ m

$F = (Gm_1 \times m_2)/r^2$
$= (6.67 \times 10^{-11} \times 70 \times 5.97 \times 10^{24})/(6.8 \times 10^6)^2$
$= 600$ N

This is just slightly less than the astronaut would weigh on the surface of the Earth.

Consolidation Questions

1   A ball is dropped from four successive heights: 1.0 m, 2.0 m, 3.0 m and 4.0 m.

   a) Calculate the time it takes to reach the ground after being dropped from these heights.

   b) Plot a graph of time to fall versus height.

   c) Explain the shape of the graph (i.e. why it curves).

2   a) An object is thrown upwards with a velocity of 12 m s^{-1}. Calculate the time it takes to reach its highest point.

   b) It then falls back down to the ground which is 2.0 m below the point from which it was thrown. How long does it take to reach the ground?

   c) What velocity does it reach just before it hits the ground?

3   The following table gives the altitude and orbital period for selected satellites.

   a) Assuming the orbits are circular and taking the radius of the Earth as 6380 km, calculate the velocity of the satellites in m s^{-1} for each of the heights.

   b) Suggest a reason why satellites need to travel more quickly when in lower orbit.

Altitude/km	Orbital period/hours
200	1.47
500	1.57
1000	1.75
5000	3.35
10 000	5.79
20 200	12
35 800	24

Table 4.2

4   A 200 kg probe is sent to Mercury.

   a) What is its weight on the surface of the planet? (You can use the information in Table 4.1 to help you.)

   Its thrusters are used to enable it to hover prior to landing. It hovers at a height of 20 m and then its thrusters are turned off and it falls to the surface.

   b) At what velocity does it hit the surface?

   c) How long did it take to fall the 20 m?

5   A tennis player hits a ball at a height of 2.5 m. The ball has an initial horizontal velocity.

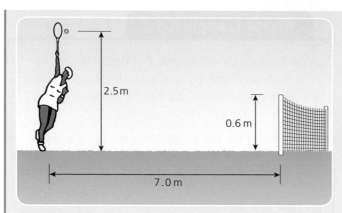

Figure 4.11

The ball just passes over the net which is 0.6 m high and 7.0 m away from her. (Neglect air friction.)

a) Calculate the time required for the ball to reach the net.

b) What was the speed of the ball as it left the racket?

6　An oil rig has to be towed to be decommissioned in the North Sea. It is towed by horizontal cables attached to two tugs as shown below.

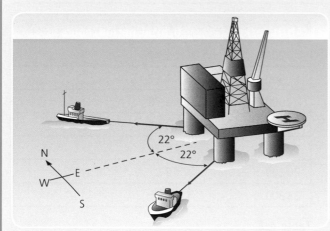

Figure 4.12

The oil rig has a mass of 25×10^6 kg and is initially at rest.

a) i) If the forces applied to the oil rig by the cables are each 1.2×10^6 N in the directions shown, what is their resultant force on the oil rig?

ii) What is the magnitude of the acceleration of the oil rig just as it moves from rest?

b) The cables continue to exert the same forces on the oil rig. The acceleration of the oil rig is continuously monitored and it is found that the acceleration decreases from its initial value. Explain this observation.

Figure 4.13

7　A model car of mass 98 g rests at the bottom of a slope as shown in Figure 4.13.

A pellet of mass 1 g is fired horizontally at a velocity of 200 m s^{-1} into the model car and remains embedded.

a) Name the type of collision.

b) To what vertical height above its initial position will the car rise, assuming it moves freely up the slope?

8*　A lift in a hotel makes a return journey from the ground floor to the top floor and then back again. The corresponding velocity–time graph is shown below.

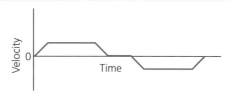

Figure 4.14

Which of the following in Figure 4.15 on the next page shows the acceleration–time graph for the same journey?

A

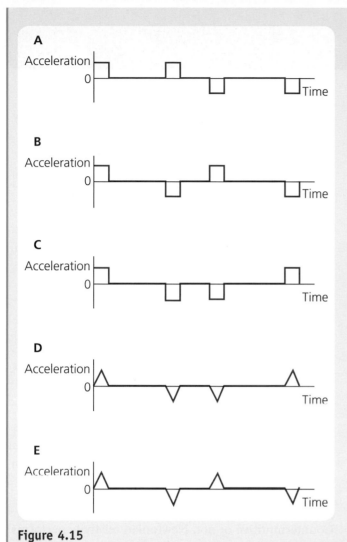

Figure 4.15

9 **a)** A long jumper devises a method for estimating the horizontal component of his velocity during a jump. His method involves first finding out how high he can jump vertically.

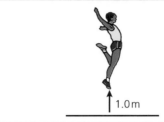

Figure 4.16

He finds that the maximum height he can jump is 1.0 m.

i) Show that his initial velocity is $4.4\,\text{m s}^{-1}$.

He now assumes that when he is long jumping, the initial vertical component of his velocity at take-off is $4.1\,\text{m s}^{-1}$.

The length of his long jump is 8.1 m.

ii) Calculate the value that he should obtain for the horizontal component of his velocity, v_H.

b) His coach tells him that, during the 8.1 m jump, his maximum height above the ground was less than 1.0 m. Ignoring air resistance, state whether his actual horizontal component of velocity was greater or less than the value stated in part **a) ii)**. You must justify your answer.

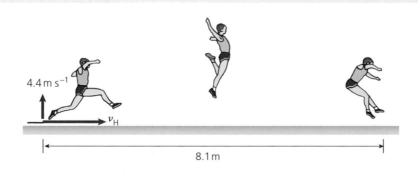

Figure 4.17

5 Special relativity

An introduction to special relativity

The earliest analyses of motion and how objects behaved when in motion were based on a principle of relativity – that is, motion relative to a fixed reference point, such as the start of a race.

The concepts of displacement, velocity and acceleration as we understand them may then be developed as:

- a position in space measured from the origin (for displacement)
- a rate of change of position (for velocity)
- a rate of change of velocity (for acceleration).

This is often referred to as **Newtonian relativity** after Isaac Newton's work which was published in the late 1600s. A further concept known as **invariance** was developed by early physicists, including both Newton and Galileo, which assumed that laws established in the laboratory held true universally and were not dependent on where they were measured or how the observer might be in motion while making the observation. This has become known as **Galilean invariance**. For example, an experiment done in a lab on land would provide the same results as one carried out in a lab in the hold of a ship at sea which was sailing steadily across the Atlantic Ocean.

With these concepts, low-velocity motions were studied and the laws of motion with which we are familiar were established. Most importantly, Newtonian relativity leads us to the concept of a frame of reference. Simply put, a frame of reference is defined by the origin or fixed reference point. As humans, we tend to put ourselves at the central, defining point of all motion. So, we might make the statement, 'the ball is moving away from me and I am stationary' even though we are standing on the surface of a planet spinning about its own axis while rotating around a star whose motion in a spiral galaxy is very rapid when observed from outside that galaxy!

All motion is relative within a frame of reference. Imagine you are travelling west along the M8 motorway to Glasgow in a car moving at 60 mph while your teacher is travelling in a car moving at the same speed but driving east towards Edinburgh. If a stationary policeman at the side of the road used a 'radar gun' to measure your speeds, he would conclude that you are both travelling within the legal speed limit of 70 mph. What speed would **you** measure with the same radar gun if you pointed it at the policeman and the teacher's car?

In the first case, you would see the policeman moving towards you (or away from you) at a speed of 60 mph – his motion relative to your own, fixed, frame of reference. In the second, you would see the teacher moving towards you (or away from you) at 120 mph – his motion relative to your own, fixed, frame of reference. This may seem counterintuitive but anyone who has sat on a train as it, or the adjacent one, moves off will realise that it is difficult to determine which one is moving and which is stationary. We measure motion relative to some fixed point and that fixed point is almost always ourselves.

Counterintuitive or not, Newtonian relativity performs well for low-velocity motion and is the basis of rocket science. It allows us to calculate the trajectory for spacecraft to travel to the Moon or even beyond.

Towards the end of the 1800s, however, a surprising consequence of James Clerk Maxwell's work on electromagnetism cast some doubt on Newtonian relativity. Maxwell, a Scottish physicist, concluded that light was an electromagnetic wave and, more importantly, that the speed of the light in a vacuum was a fixed, constant value which could be determined theoretically from the values of two fundamental constants related to electric and magnetic fields. The **permittivity of free space** (ε_0) and the **permeability of free space** (μ_0) are constants of proportionality which may be derived from experiment. Maxwell's theoretical work showed that the speed of light, c, was given by:

$$c = \frac{1}{\sqrt{(\varepsilon_0 \times \mu_0)}}$$

Einstein took this result and worked it through to its conclusion based on two postulates:

- The speed of light is a constant value in every frame of reference.

- Galilean invariance holds true for all frames of reference.

Einstein performed thought experiments to test the consequences of his theories so this is a good place to start.

Imagine you are again travelling west on the M8 but this time you are travelling at $0.5c$ (in other words, half the speed of light). The traffic policeman is still there and uses his radar gun to measure not your speed but the speed of light waves emitted from your headlamps.

Newtonian relativity predicts that the light should arrive at the policeman at the combined speed of your motion and the speed of the light wave – in this case $1.5c$. Maxwell's work predicts that the speed of light is a universal constant regardless of relative motion and that the light always arrives at the policeman at the same speed, c.

Maxwell's prediction is the correct one and as a result, Einstein arrived at his Theory of Special Relativity – because the speed of light is a universal constant, the nature of space and time for a moving object are changed relative to a stationary observer. From this it can be shown that time is measured according to the relative velocity of the frame of reference in which it is measured! That is, the results of length and time measurements made by a stationary observer and a moving observer would be different. The stationary observer would record time passing more quickly than the moving observer and would measure distances as longer than those measured by the moving observer.

The ratio of distance and time is used to calculate speed and from this the constant speed of light ensures there will be proportional changes in the times and distances measured, in other words, time dilates at the same rate as lengths contract.

Let us consider how Einstein arrived at these startling conclusions by repeating the thought experiments he used.

Time dilation

Both thought experiments involve a 'light clock' which is simply a pair of parallel mirrors with a light beam reflecting between them. The clock measures the time for the beam to travel between the mirrors.

Imagine a beam of light bouncing vertically between two mirrors arranged as shown in Figure 5.1.

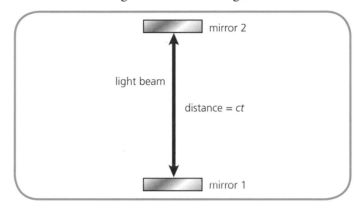

Figure 5.1 A 'light clock'

An observer would see the light beam travelling between the two mirrors take a time t such that the distance between the mirrors is given by ct.

Now imagine the two mirrors are travelling at velocity v. An observer moving with the mirrors would discern no difference. A stationary observer would see the light beam moving diagonally as a result of the relative motion. Figure 5.2 shows the light apparently travelling a greater distance which is the vector combination of vt' (mirror displacement) and ct (mirror separation).

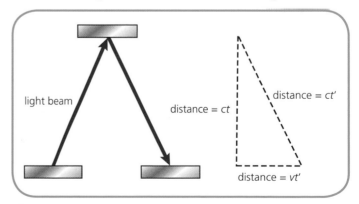

Figure 5.2

From the figure, Pythagoras gives:

$$(ct')^2 = (ct)^2 + (vt')^2$$
$$t'^2 = t^2 + (vt'/c)^2$$
$$t'^2(1 - (v/c)^2) = t^2$$

Quantity	Name	Unit	Symbol
t'	observed time (stationary observer)	seconds	s
t	observed time (moving observer)	seconds	s
v	velocity of object observed	metres per second	$m\,s^{-1}$
c	speed of light	metres per second	$m\,s^{-1}$

Table 5.1

$$t' = t \times \frac{1}{\sqrt{1 - (\frac{v}{c})^2}}$$

Table 5.1 shows what each of the variables of the formula represent. The startling result, based on the postulate that the speed of light is constant when measured by a moving or stationary observer, is that the times they measure must be different by a scaling factor, called the **Lorentz factor**, and given the symbol γ (Greek letter gamma).

$$\gamma = \frac{1}{\sqrt{1 - (\frac{v}{c})^2}}$$

which depends on the size of the velocity, v, of the moving object relative to the stationary observer.

So, $t' = \gamma t$. In other words, the observed time, t', elapsing is longer than the time experienced in the moving frame of reference. Therefore when measuring or predicting the motions of fast-moving objects, stationary observers must correct for the time effects of **time dilation** in the moving frame.

A graph of γ versus v/c is shown in Figure 5.3.

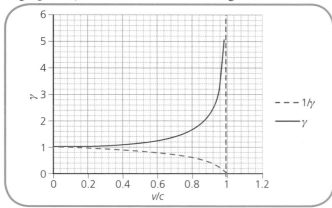

Figure 5.3 A graphical representation of γ versus v/c

Note that for low values of v relative to c then $\gamma \approx 1$ and so the special relativity corrections collapse to give the traditional Newtonian relativity results. However, as v increases beyond 10% or so of c, then the value of γ begins to rise to a significant level and special relativity corrections need to be taken into account.

Length contraction

In a similar argument, the effect of high-speed relative motion on length measurements may be derived. The measurement of length of a fast-moving object by a stationary observer requires that two locations – the start and end positions of the object – are measured simultaneously. Given that time is measured according to the relative velocity of the frame of reference in which it is measured, this simultaneous measurement is difficult to achieve.

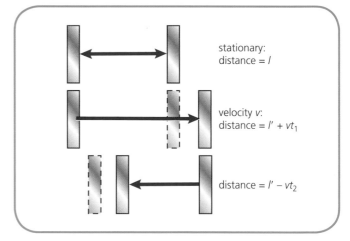

Figure 5.4

On this occasion consider a light clock positioned parallel to the direction of a moving train. For an observer inside the train, the time required to complete the round trip between the mirrors is:

$$t = \frac{2l}{c}$$

To the observer outside the train, from the above the time t to make a round trip appears to be:

$$t' = t \times \frac{1}{\sqrt{1 - (\frac{v}{c})^2}}$$

due to time dilation. This gives

$$t' = \frac{(\frac{2l}{c})}{\sqrt{1 - (\frac{v}{c})^2}} \quad \text{(Equation 1)}$$

A stationary observer outside the train also sees the mirrors constantly moving to the right.

Let t_1 be the time for the photon to go from the left mirror to the right mirror. The right mirror then travels away a distance vt_1 before the photon hits it. Hence, the photon travels a distance $d_1 = ct_1 = vt_1 + l'$

$$t_1 = \frac{l'}{(c - v)}$$

On the return pass, let t_2 be the time for the photon to go from the right mirror to the left mirror. The left mirror then approaches a distance vt_2 before the photon hits it. Hence, the photon travels a distance

$$d_2 = ct_2 = l' - vt_2$$

Thus,
$$t_2 = \frac{l'}{(c + v)}$$

The total time measured by the stationary observer is

$$t' = t_1 + t_2 = \frac{l'}{(c - v)} + \frac{l'}{(c + v)}$$

So,
$$t' = \frac{\left(\frac{2l'}{c}\right)}{1 - \left(\frac{v}{c}\right)^2} \text{ (Equation 2)}$$

Equations 1 and 2 are for the same time, t'. Equating these two gives

$$\frac{\left(\frac{2l'}{c}\right)}{1 - \left(\frac{v}{c}\right)^2} = \frac{\left(\frac{2l}{c}\right)}{\sqrt{1 - \left(\frac{v}{c}\right)^2}}$$

So,
$$l' = l\sqrt{1 - \left(\frac{v}{c}\right)^2}$$

or
$$l' = \frac{l}{\gamma}$$

Table 5.2 shows what each of the elements of the formula represent.

In other words, the length measured by a stationary observer, l', is shorter than the length measured by an observer in the moving frame of reference.

Therefore when measuring or predicting the motions of fast-moving objects, stationary observers must correct for the time effects of **length contraction** in the moving frame. Note that it is only the length measured in the direction of travel which is contracted, so a 3D object would be distorted only in one dimension.

If you remain unconvinced by these theoretical conclusions derived from the two postulates of special relativity then do not worry. When they were proposed, there were many who were similarly sceptical; some scientists could not or would not accept the conclusions. Sir Arthur Eddington, a British scientist, was one of the few who understood special relativity at the time. It is said that on being told during an interview that he was one of only three people who understood Einstein's theories, Eddington replied, after a long pause, 'Who is the third person?'

Experimental proof of both effects has since confirmed that Einstein's special relativity effects are real and observable.

Muon detection and time dilation

Muons are naturally occurring sub-atomic particles which may result from cosmic radiations impacting the Earth's atmosphere. They are unstable particles with a half-life of around 1.56 μs when measured in a laboratory. It is possible to make high-altitude measurements and count how many muons are detected per second. Equally, the same measurements can be made at ground level. When this is done, the results can only be explained using special relativity.

Imagine that the high-altitude observations made at 10 km counted 1 million muons. Given the half-life of 1.56 μs and the muon velocity of around 0.98 c, how many should reach the Earth's surface? Classically, the time for the journey is given by

$$t = 10\,000/0.98\,c = 34 \times 10^{-6}\,\text{s}$$

Quantity	Name	Unit	Symbol
l'	observed length (stationary observer)	metre	m
l	observed length (moving observer)	metre	m
v	velocity of object observed	metres per second	m s^{-1}
c	speed of light	metres per second	m s^{-1}

Table 5.2 Variables in the length contraction equation

This is around 22 half-lives which gives a survival rate of around 3×10^{-7} (i.e. 2^{-22}), or 0.3 in every million.

In reality, measurements at the Earth's surface would count around 49 000 muons for every million measured at an altitude of 10 km. This is clearly different from 0.3!

Using special relativity, as the muons travel at almost the speed of light, the Lorentz factor is given by

$$\gamma = \frac{1}{\sqrt{1 - \left(\frac{v}{c}\right)^2}}$$

In this case, $\gamma = 5$.

This means $t' = \gamma t$, that is the Earth observer's measurement of the muons' journey time, t', is five times greater than that experienced by the muon. In other words, the muons' 'time clock' runs slow, so the stationary observer measures its half-life as 7.8 μs.

Alternatively, using $l' = l/\gamma$, the length of the journey experienced by the muon, l', is shorter by a factor of five. In other words, length contraction means the muon 'sees' the distance travelled as only 2 km.

Either correction, time dilation or length contraction, allows the journey to be shown to be the equivalent of around 4.4 half-lives, which generates survival rates in agreement with experiment.

Table 5.3 summarises the results of the two approaches.

Fast-moving clock corrections

Experiments and everyday use of fast-moving clocks provide further evidence for the validity of this relativistic theory.

In 1971, Hafele and Keating used accurate atomic clocks to test Einstein's theories of relativity. They first synchronised all clocks with the US Naval Observatory atomic clock, then had them flown twice around the world and finally compared their reading of the time with the US naval clock again.

The clocks had indeed slowed as a result of their journey eastward and gained time after westward travel in accordance with Einstein's predictions. (Note that this experiment tested both special and general relativity as the latter requires corrections for the effects of the Earth's gravitational field.)

One practical consequence of this is that the time signals generated by the atomic clocks travelling on board global positioning satellites (GPS) must be compensated to account for the effects of relativity as the satellites are travelling at high speed and will 'lose' roughly 7200 ns/day due to special relativity time dilation. When such satellites were first launched in 1997, there was still doubt as to whether the clocks' operation would be affected by relativity and the corrections were not enabled. After around 20 days, the time on the satellite clocks was sufficiently different from that at the stationary ground stations to confirm the predicted effects of Einstein's theories of relativity and the correction circuits were turned on. Again, these corrections were for both special and general relativity.

	Relativistic		Non-relativistic
	Muon	Ground	
Distance	2 km	10 km	10 km
Time	6.8 μs	34 μs	34 μs
Observed half-life	1.56 μs	7.8 μs	1.56 μs
Number of half-lives	4.36	4.36	21.8
Ground count	49 000	49 000	3

Table 5.3 Muon journey summary

Consolidation Questions

1 A train travels through a station at a constant velocity of $+10\,\text{m}\,\text{s}^{-1}$. Anne is sitting on the train and Bob is standing on the platform. Meanwhile Charles is on the train, walking in the same direction as the train is travelling as he makes his way to the buffet car.

   a) Anne estimates Charles' velocity as $+1\,\text{m}\,\text{s}^{-1}$. What does Bob estimate as Charles' velocity?

   b) After 10 seconds, how far has Charles moved relative to Anne?

   c) After 10 seconds, how far has Charles moved relative to Bob?

2 Maxwell derived the equation for the speed of light (c) shown below:

$$c = \frac{1}{\sqrt{\varepsilon_0 \times \mu_0}}$$

Look up the values of ε_0 (permittivity of free space) and μ_0 (permeability of free space) and use the equation to calculate a value for c.

3 For Einstein's theory of Special Relativity:

   a) What are the two postulates of the theory?

   b) What does Galilean invariance mean?

   c) For two spacecraft moving apart at $0.75\,c$, the separation between them is measured as increasing at a rate of $1.5\,c$.

   i) Explain why this measurement is consistent with Einstein's postulates.

   ii) Calculate the speed of one of the spacecraft as measured by an observer in the other.

4 Muons are sub-atomic particles which may be formed as the result of collisions between cosmic rays and gas molecules in the Earth's upper atmosphere. They are inherently unstable with a half-life of $1.56 \times 10^{-6}\,\text{s}$, decaying to produce an electron. Muons produced in the upper atmosphere are fast moving with speeds of around $0.98\,c$ relative to an observer on Earth.

   a) The half-life of the muon in its frame of reference is $1.56 \times 10^{-6}\,\text{s}$. What value for the half-life will be measured by observers on the Earth's surface, assuming the muons are produced at an altitude of $10\,\text{km}$?

   b) For every million (10^6) muons, how many will be measured arriving at a detector on the Earth's surface?

   c) What distance would be measured by an observer travelling with the muons?

6 The expanding Universe

The Doppler Effect and red shift of galaxies

Imagine yourself in the stand at a Formula 1 Grand Prix as the cars drive past. The predominant note from the engine noise does not change but what you hear as a stationary observer does. The high note from the engine as the car comes towards you appears to change to a lower note as it drives away.

This is another classical relativistic effect. The sound wave being produced is at a constant frequency and a fixed wavelength but the effect of the relative motion of the wave emitter and detector (the engine and your ears, respectively) is to cause an apparent change in the frequency of the detected wave. In the case of the Formula 1 cars, the engine note appears to be at a higher frequency on approach and lower frequency on moving away from the observer. This phenomenon is known as the **Doppler Effect** and is observed for all waves, including sound and light.

The Doppler Effect on frequency results from a compression or elongation of the effective distance between wavefronts as a result of the relative motion of the wave source and detector.

As the wave source is moving towards the detector, each successive wavefront is emitted from a distance slightly closer than the previous one. Accordingly, each wave arrives a little more quickly (earlier) than its predecessor as the wave speed and relative motion combine. This causes the frequency, in other words the number of waves arriving at the observer per second, to be increased.

As the wave source is moving away from the detector, each successive wavefront is emitted from a distance slightly further than the previous one. Accordingly, each wave arrives a little more slowly (later) than its predecessor as the wave speed and relative motion are in opposite directions. This causes the frequency, in other words the number of waves arriving at the observer per second, to be decreased.

The equation for the observed frequency measured by the stationary observer is given as

$$f_o = f_s \times \frac{v}{(v \pm v_s)}$$

Table 6.1 shows what each of the terms of the formula represent. (Note that the ± must be selected for the relative motion of the emitter towards (−) or away from (+) the observer.)

Given the form of the equation, the Doppler Effect is observable when the velocity of the source, v_s, is significant compared to the wave velocity, v.

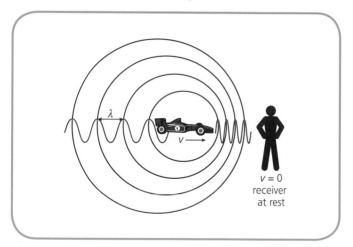

Figure 6.1 The Doppler Effect

Quantity	Name	Unit	Symbol
f_o	observed frequency	hertz	Hz
f_s	source frequency	hertz	Hz
v	velocity of wave	metres per second	$m\,s^{-1}$
v_s	velocity of source	metres per second	$m\,s^{-1}$

Table 6.1 Variables in the Doppler equation

As noted, the Doppler Effect is also observed for moving objects which emit light. It is possible to see a frequency change on the light detected when the speed of movement is sufficiently large. This is the case for the light emitted by stars moving relative to us on Earth. The light emitted or absorbed by an object has a characteristic series of lines or wavelengths which relate to the atomic structure of the chemical elements which make up the object. As white light emitted by a star passes through the outer gaseous layer of the star itself, particular wavelengths will be absorbed by the constituent chemical elements of the star and these missing wavelengths, so-called **Fraunhofer lines**, may be detected by spectral analysis here on Earth.

Fraunhofer was a German physicist and he published a work in 1814 describing over 500 'missing' wavelengths from the spectrum of our Sun. In fact, these lines subsequently provided a chemical fingerprint of the constituents of the Sun's atmosphere and have been used to help analyse other stars as telescopes improved.

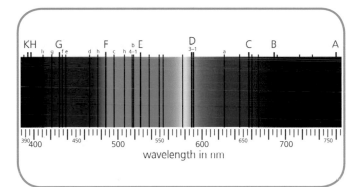

Figure 6.2 Solar Fraunhofer lines

Spectral analysis of the light from distant stars shows similarly spaced lines. In other words, stars are made up from the same chemical elements, but the absolute line positions appear to have been shifted up or down the spectrum. This is a result of the motion of the star relative to us and is direct evidence of the Doppler Effect on the light waves emitted from the star. It is usual to speak of a Doppler shift in the wavelength, as opposed to the frequency, of the light as this appears to be a more convenient unit of measurement.

If the star is moving away from us then the spectrum of Fraunhofer lines is said to be 'red shifted', in other words moved to a longer wavelength value towards the red end of the spectrum. If the star is moving towards us then the spectrum of Fraunhofer lines is said to be 'blue shifted', in other words moved to a shorter wavelength value towards the blue end of the spectrum.

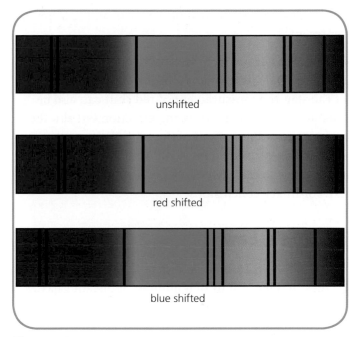

Figure 6.3 Shifted Fraunhofer lines

The physical quantity called **red shift** is given by the equation

$$z = \frac{(\lambda_{observed} - \lambda_{rest})}{\lambda_{rest}}$$

Table 6.2 shows what each of the terms of the formula represent.

Quantity	Name	Unit	Symbol
z	red shift	none	none
$\lambda_{observed}$	observed wavelength	metre	m
λ_{rest}	rest wavelength	metre	m

Table 6.2 Variables in the red shift equation 1

(Note that z will be positive for stars moving away from us and negative for stars moving towards us.)

Given our discussion of relativity in the previous chapters, it follows that for stars and galaxies moving relative to Earth with very high velocities, corrections need to be made for relativistic effects. Those corrections are beyond the level of this course.

For lower, non-relativistic velocity galaxies, the red shift is also given by

$$z = \frac{v}{c}$$

Table 6.3 shows what each of the terms in the formula represent.

The analysis of star light is capable of yielding a great many results. Quite apart from the chemical make-up of the star, the measurement of red shift can also be used to determine not only the direction but also the magnitude of the motion of the star relative to the Earth. This is clearly a powerful technique that when applied in the early twentieth century led to some truly extraordinary conclusions.

Hubble's Law

In 1929, Edwin Hubble published the results of his studies of the red shift of a number of distant galaxies based on his work at the, then, state-of-the-art 2.5 m reflector telescope at Mount Wilson in California. Based on his observations of 24 very bright distant stars (Cepheid variables), he was able to state a new relationship:

$$v = H_0 d$$

Table 6.4 shows what each of the terms in the formula represent.

Hubble found that all stars were moving away from Earth and that the further stars were moving away at a greater velocity than closer stars. This discovery that

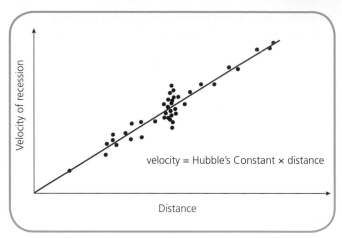

Figure 6.4 Graph of Hubble's Law

all stars were 'moving away' from us was considered profound.

Hubble's original data based on the Cepheids visible to him gave a value of $H_0 = 1.62 \times 10^{-17}\,\text{s}^{-1}$, based on the line of best fit on the graph above. Subsequent work using more sensitive telescopes on Earth and, latterly, ones in orbit beyond the atmosphere have allowed astronomers to look at very, very distant stars and galaxies. This increase in the range of distances to the galaxies being examined has significantly improved the accuracy of the straight-line fit and leads to the current estimate of $H_0 = 2.34 \times 10^{-18}\,\text{s}^{-1}$.

A possible explanation for this is that there was one event that created and emitted all the matter in the Universe from a single point. This matter spread out creating the Universe as we know it. Objects moving quickly are at the extreme ends of the Universe and

Quantity	Name	Unit	Symbol
z	red shift	none	none
v	relative velocity of star	metres per second	m s^{-1}
c	speed of light	metres per second	m s^{-1}

Table 6.3 Variables in the red shift equation 2

Quantity	Name	Unit	Symbol
H_0	Hubble constant	per second	s^{-1}
v	relative velocity of star	metres per second	m s^{-1}
d	distance to the star	metre	m

Table 6.4 Variables in the Hubble's Law equation

objects moving less quickly are nearer the 'centre'. This would mean that observers like ourselves would see distant objects moving away from us and that we would be moving away from slower objects nearer the centre. From our observations all galaxies appear to be moving away from us.

It should be noted that Hubble's equation is an equation of direct proportion – the further away the star (galaxy) is from Earth, the faster it is moving away from us. This recession velocity increase with distance was unpredicted but subsequent measurements reliant on improving telescope technology have verified the original form of this law.

The Hubble constant has a further use. The law suggests that the Universe is expanding and if the rate of expansion is assumed to be fixed, then the value of the Hubble constant may be used to determine the age of the Universe. Rather than running the clock forward to measure the distance to a distant star, Hubble's Law can be rearranged to determine how long the star has been moving. In other words, when did the Universe begin?

$$\text{time} = \frac{d}{v}$$

but Hubble's Law gives

$$\frac{d}{v} = \frac{1}{H_0}$$

therefore

$$\text{time} = \frac{1}{H_0}$$

So $1/H_0$ yields the age of the Universe. From the original measures in 1929 to the currently accepted values, the use of H_0 has resulted in a figure for the age of the Universe of the order of 2 to 13.7 billion years, respectively. The wide range has resulted from the very significant improvements in the accuracy of astronomical measurements made in the second half of the twentieth century. All recent measures place the age of the Universe at around 13 billion years. Spectral analysis of light gathered by a telescope is a powerful technique indeed!

(It should be noted that the SQA course requires the use of SI units, hence the non-traditional use of s^{-1} and $m\,s^{-1}$ in this section. More usually H_0 is quoted in units of km/s/Mparsec (kilometres per second per megaparsec) with velocity in km/s and distance in Mparsec. While this changes the numerical value of H_0 used here compared to elsewhere, the principle of how to determine Universe age remains the same.)

Evidence for the expanding Universe

Hubble's Law tells us that the Universe is expanding and the Hubble constant can be used to estimate the minimum age of the Universe, assuming that the rate of expansion has been constant since its inception. A consequence of this observation is that the Universe originated from a single fixed point where all the matter and energy of the Universe was concentrated into an infinitesimally small volume. This discontinuity in space was the very beginning of all matter – galaxies, stars, planets and even the life forms which inhabit those worlds.

The application of Newtonian or classical physics to this matter tells us that the velocities of the exploded fragments from the original super-explosion must be constant in the absence of unbalanced forces. The unbalanced force which must be acting between these astronomical objects is gravitational attraction. Gravitational attraction is due to the mass of matter.

This leads to the idea that the total mass in the Universe will determine whether the rate of expansion is indeed constant or is decreasing over time as all of the matter spread out over the Universe interacts gravitationally and so acts as an 'attractive' brake on the expansion rate.

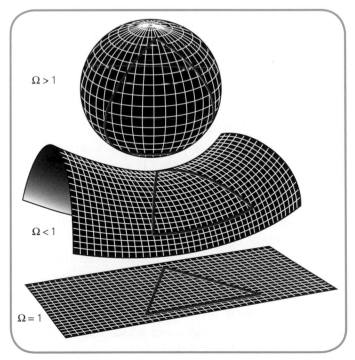

Figure 6.5 The three 'geometries of space'

Since the latter part of the twentieth century, physicists have been wrestling with measuring the expansion rate and estimating the total mass of the Universe. (The ultimate fate of the Universe is related to its density.) They do this by comparing actual density to a critical density value (which is proportional to H_0^2) and calculating the ratio known as omega (Ω), where

$$\Omega = \text{actual density/critical density}$$

There are three possibilities:

- $\Omega > 1$: the actual density > critical and so the expansion will eventually reverse due to gravitational attraction and the Universe will then collapse in on itself – the 'Big Crunch'.

- $\Omega < 1$: the actual density < critical and so the expansion will continue and may even accelerate, possibly leading to the 'Big Rip' where the Universe is torn apart by the expansion.

- $\Omega = 1$: the actual density = critical and so the Universe is infinite and will continue to expand at a constant rate for ever.

These give rise to three 'geometries of space' represented by Figure 6.5 showing closed, open and flat for the possible values of Ω.

The mass of the Universe is estimated by measuring the speed of the radial motions of stars and galaxies. Measurements of these give us an indication of how galaxies are moving and interacting and this gives an indication of their masses.

Once again, physical relationships determined in the laboratory may be applied to huge objects many light years distant from us to establish their physical characteristics. The force which causes a celestial object to rotate around another is gravitational attraction which depends on the mass of the objects, where greater central mass results in greater rotational speeds. Measurements of these galaxies have shown that the rates of rotation are too large for the matter which may be observed. This has led to an astounding conclusion – a new type of matter exists in our Universe. This type of matter does not interact with the electromagnetic spectrum and so is invisible to astronomers using telescopes which depend on gathering reflected or emitted electromagnetic waves (from radio to gamma) to observe distant objects.

As a result this new matter has been called **dark matter**. Current estimates are that 22% of the total matter and energy of the Universe is made up of this type of invisible matter.

The introduction of huge, high-altitude and extra-terrestrial telescopes has allowed for the observation of very distant objects. For example, the Supernova Cosmology Project has looked at Type Ia supernovae in galaxies with a red shift of $z = 0.51$ which equates to looking back to about 40% of the age of the Universe. When the data from these distant galaxies and stars are plotted on a Hubble diagram (as can be seen at http://www.eso.org/~bleibund/papers/EPN/epn_fig2.jpg), the relationship is found to be non-linear at large distances. In other words, the most distant objects are accelerating away from us!

If that is the case then the additional kinetic energy gained by these objects must be derived from some source that we cannot fully explain at present. This has led to a proposal of the existence of a type of

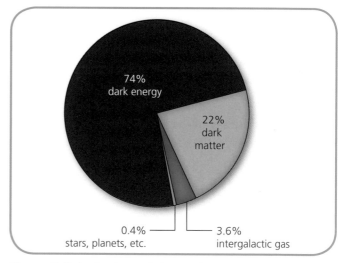

Figure 6.6 A pie chart of the total matter and energy in the Universe

Type	Percentage of total mass and energy/%
stars, planets, etc.	0.4
intergalactic gas	3.6
dark matter	22.0
dark energy	74.0

Table 6.5 Energy and matter distribution in the Universe

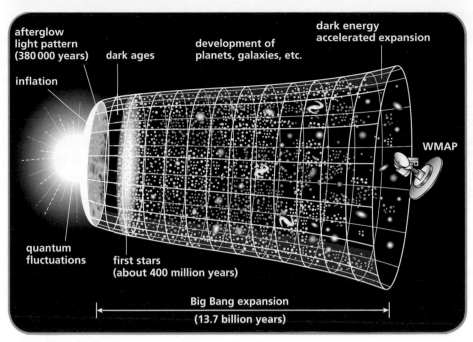

Figure 6.7 The expansion of the Universe

dark energy. Similar to dark matter, this type of energy is not detectable by the types of sensors we use to record electromagnetic energy radiations. It is dark energy which we believe is responsible for this accelerating expansion rate. Current estimates are that approximately 74% of the total matter and energy of the Universe is this type of 'invisible' or difficult-to-detect energy. Including measurements for dark energy suggests that $\Omega < 1$ but not by much.

The startling inclusion of these new types of energy and matter means that the matter which makes up all planets, stars and life forms accounts for a diminishingly small part (4%) of the total matter and energy of the Universe. This is shown as a pie chart in Figure 6.6.

The current picture of Universe expansion is depicted in the image shown in Figure 6.7.

This timeline is a representation of the evolution of the Universe over 13.7 billion years. The far left depicts the earliest moment we can now probe, when a period of 'inflation' produced a burst of exponential growth in the Universe. (Size is depicted by the vertical extent of the grid in this graphic.)

For the next several billion years, the expansion of the Universe gradually slowed down as the matter in the Universe pulled on itself via gravity. More recently, the expansion has begun to speed up again as the repulsive effects of dark energy have come to dominate the expansion of the Universe.

The afterglow light seen by WMAP (the Wilkinson Microwave Anisotropy Probe) was emitted about 380 000 years after inflation and has traversed the Universe largely unimpeded since then. (WMAP is a space probe placed at a special position which allows it to measure radiation from space extremely accurately.) The conditions of earlier times are imprinted on this light; it also forms a backlight for later developments of the Universe.

Consolidation Questions

In the following questions, take the speed of sound in air to be $340\,\text{m s}^{-1}$.

1   A train is moving with constant speed of $8.0\,\text{m s}^{-1}$ along a straight section of track towards an observer standing (dangerously) close to the track edge. The driver sounds the train horn which emits a single note of $277\,\text{Hz}$ (middle C). What three frequencies will be heard by the observer as the train completes its journey past him?

2   A stationary source emits a sound of frequency $440\,\text{Hz}$ which is detected by a high-speed observer who measures the frequency to be $480\,\text{Hz}$.

   a) What is the speed of the observer?

   b) What wavelength would be measured by
    i) someone stationary beside the source?
    ii) someone moving with the observer?

3   Ultrasound may be used in medicine to determine the speed of blood flow. Using a $7.0\,\text{MHz}$ ultrasound source, reflected sound waves are measured with frequency shift of $2.5\,\text{kHz}$. Given that the speed of sound in tissue is $1540\,\text{m s}^{-1}$, what is the speed of the blood cells which reflect the sound signals?

4   One of the characteristic wavelengths of the Balmer series in the hydrogen spectrum is $656\,\text{nm}$. Light from a distant galaxy is detected on Earth and this emission line is measured to be $633\,\text{nm}$.

   a) Is the galaxy approaching or moving away from Earth?

   b) What is the size of the red shift for this galaxy?

   c) What is the speed of the galaxy relative to Earth?

   d) Estimate the distance from the Earth to the galaxy.

5   State Hubble's Law. Make sure that you explain the meanings and units of any symbols you may have used. What data led Hubble to his conclusion?

6   Hubble's constant, H_0, is usually quoted in units of $\text{km s}^{-1}\,\text{Mpc}^{-1}$. What are the meanings of these unit symbols? Why would they be chosen rather than SI units?

7   Hubble's original value of H_0 was $500\,\text{km s}^{-1}\,\text{Mpc}^{-1}$. Currently the value is estimated as $72.5\,\text{km s}^{-1}\,\text{Mpc}^{-1}$.

   a) Use these two values to estimate an age for the Universe.

   b) What are the units for your answers?

   c) Why is there such a difference between the original and current values?

7 The Big Bang Theory

The temperatures of stellar objects

We observe stars on Earth using the light and other electromagnetic radiations they emit. All frequencies of electromagnetic radiation are used in astronomy from low-frequency radio waves to very-high-frequency X-ray and gamma radiation. The energy E of the radiation is directly proportional to its frequency as shown by the Einstein–Planck equation ($E = hf$, where h is Planck's constant and f is frequency) and so by measuring the relative amount of each type of radiation a star emits it is possible to make deductions about the thermal energy, in other words, the temperature of the star. It is more usual to plot radiation intensity versus wavelength when examining the radiation emitted by a star. An example of such a graph is shown in Figure 7.1.

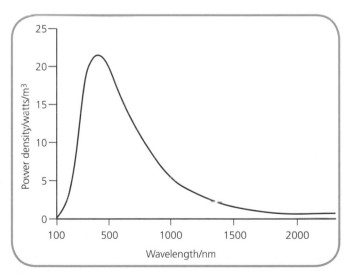

Figure 7.1 A graph of radiation intensity against wavelength

Note that the graph is a curve with intensity rising from a very low value at short wavelengths to a peak at some intermediate wavelength value and then falling to a low value as wavelength increases. This is a typical curve not only for a star but for any hot object emitting electromagnetic radiation – a so-called 'blackbody radiator'.

The shape of this curve has been studied experimentally since the mid-1800s. It is known that the shape of the curve and the wavelength position of the peak intensity (the peak wavelength) depend directly on the temperature of the object.

- High temperatures correspond to high energies and are therefore associated with lower peak wavelength values.

- For similar stars (mass, surface area, etc.) hotter stars will emit more energy than cooler stars and so the area under the curve will be greater.

There is an experimental law, known as Wien's Displacement Law, which states that the product of the peak wavelength and the temperature is found to be a constant, in other words:

$$\lambda_{peak} \times T = 2.898 \times 10^{-3}\,\text{m K}$$

Table 7.1 shows the meaning of each element of this equation.

Quantity	Name	Unit	Symbol
2.898×10^{-3}	constant	m K	none
λ_{peak}	peak wavelength	metre	m
T	stellar temperature	kelvin	K

Table 7.1 Variables in the Wien's Displacement Law equation

(Note: this equation does not form part of the SQA course.)

This means we now have a stellar thermometer!

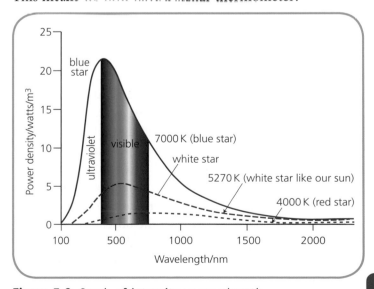

Figure 7.2 Graph of intensity vs wavelength

Spectral type	Surface temperature/K	Chemistry
O	> 25 000	H; He–I; He–II
B	10 000–25 000	H; He–I; He–II absent
A	7 500–10 000	H; Ca–II; He–I and He–II absent
F	6 000–7 500	H; metals (Ca–II, Fe, etc.)
G	5 000–6 000	H; metals; some molecular species
K	3 500–5 000	metals; some molecular species
M	< 3 500	metals; molecular species (TiO)

Table 7.2 The Harvard spectral classification

From the comfort of the laboratory or observatory it is possible to measure the surface temperature of a star simply by measuring the intensities of the various wavelengths of light it emits. Figure 7.2 shows the radiation curves for various stellar temperatures.

Note that the peak wavelength moves to the left of the graph and the area under the curve becomes larger as the star surface temperature increases.

This allows for the classification of stars based on their surface temperatures, which is intimately related to the physical processes occurring during the star's life cycle. Stars are classified using the letters O, B, A, F, G, K and M. Table 7.2 shows the Harvard spectral classification with the temperature and chemical make-up associated with each class.

Evidence for the Big Bang

Published in the 1920s, Hubble's Law was the first inkling that a cosmic starting point could be identified. The present-day expansion of the Universe must mean that in earlier times the Universe was smaller and that if we go back to even earlier times the Universe was much smaller still. Taking this to its logical conclusion, if the Universe is currently expanding then at some earlier time it must have been compressed into a single fixed point where all matter and energy of the Universe was concentrated into an infinitesimally small volume. This is/was the contentious theory derived from the Hubble diagram but other than the observed expansion of the Universe, what additional evidence might support the idea that all matter and energy originated from a single point or event?

Large-scale homogeneity

When we say that the Universe has large-scale homogeneity, we mean that on the cosmological scale (hundreds of millions of light years), it appears to be the same in all directions. Homogeneity means that there is a uniform distribution of matter regardless of where you are in the Universe. This is what you would expect if there was an initial 'explosion' – matter would be distributed equally in all directions. The fact that this is true is consistent with a Big Bang expansion from an initial hot, dense state.

Figure 7.3 shows an image called the Hubble Deep Field (HDF), taken with NASA's Hubble Space Telescope. It is taken at extremely high magnification pointing at a minute section of the sky but shows an incredible number of galaxies never before visible and provides evidence for the homogeneity of the Universe.

Figure 7.3 A Hubble Deep Field image

The abundances of light elements

As the Universe expanded and cooled, the earliest matter (baryons, protons and neutrons) condensed out of the energetic starting point but the temperatures were still high enough for them to exist as free particles. During the period known as Big Bang nucleosynthesis, which lasted from around 3–20 minutes after the Big Bang, these particles underwent nuclear fusion to create light chemical elements.

Consideration of the temperature and conditions at this time leads us to predict that these were mostly nuclei of helium (^{3}He and ^{4}He), deuterium (^{2}H), lithium and beryllium.

As these newly formed elemental nuclei were not constrained, as within a stellar core for example, and due to the continued cooling of the Universe, they did not continue to fuse to form any heavier elements. They were allowed to spread freely throughout the Universe.

Current-day determination of the relative abundances of these chemical elements is in agreement with the ratios predicted by this theory of formation.

Existence of cosmic microwave background radiation (CMBR)

During the nucleosynthesis period stable atoms were not created because the energy density was too high for electrons to be bound to a nucleus. The physical conditions in existence at this time led to a hot, dense plasma with free electrons and protons moving independently at a high speed. This meant photons of light could not pass easily through the plasma without either scattering off these charged particles or being absorbed and freeing any electron bound to a proton. With further expansion and cooling of the Universe (about 400 000 years after the Big Bang) the temperature dropped sufficiently for neutral atoms to form.

From this point, only particular photon energies could interact with the newly formed matter and so the other photons could now travel freely across all space – the Universe had become transparent.

With the subsequent Universe expansion, the wavelengths of these photons were similarly expanded by a factor of around 1000. This radiation exists today

as the **cosmic microwave background radiation** and should be characterised by the following two attributes:

- It should be uniformly distributed across the Universe.

- It should have a radiation intensity curve like a blackbody radiator with a predictable peak wavelength.

Originally proposed in the 1940s, the CMBR was finally measured in 1965 by Penzias and Wilson. They did so accidentally while working on the use of microwaves for telecommunication. They were frustrated at their inability to remove a persistent background noise from their signals despite all attempts to do so. It transpired that the 'noise' was in effect the Universe cooling down and they were receiving microwaves as part of a blackbody spectrum associated with a body whose temperature was around 3 K. Penzias and Wilson received a Nobel Prize for their accidental discovery.

Many more experimental measurements followed this initial work with the COBE satellite (COsmic Background Explorer) making detailed observations from beyond the thermal interference of the atmosphere. The blackbody spectrum from the COBE measurements is shown in Figure 7.4.

The astounding level of agreement with CMBR measurement and the Big Bang theory prediction is real evidence to support the theory. Regardless of some variation in the value of the predicted background

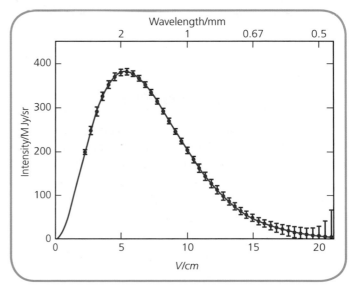

Figure 7.4 Graph of blackbody spectrum measurements

temperature, there is no other mechanism that gives rise to a uniform, long-wavelength background radiation in the Universe. Other models might predict some absorption/radiation-based origin for microwave wavelengths but none give rise to the uniformity or type of blackbody spectrum seen in the CMBR.

Fine structure of the CMBR

More detailed measurements with very high contrast applied to the images of the CMBR show a level of structure or anisotropy which was predicted by the refined models of the Big Bang Theory and relates to the physical processes which were occurring around

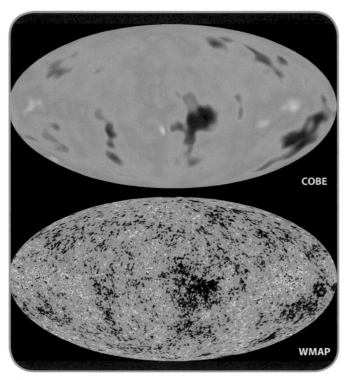

Figure 7.5 Comparing COBE and WMAP

the time of the matter–radiation equality in the Universe expansion after the Big Bang. At this point, the theory postulates that dark matter clumps would be formed due to gravitational attraction. These would gravitationally attract the matter–radiation plasma and so any fine structure would be 'frozen' in the CMBR after subsequent expansion in a similar way to the method in which biologists stain samples before viewing with a microscope. If the Big Bang Theory is correct, such fine structure should be visible today.

In 2001, the WMAP (Wilkinson Microwave Anisotropy Probe) was launched specifically to look for fluctuations in the CMBR. Improved by the experiences of COBE, instead of orbiting the Earth it was sent to orbit the Sun at a stable point behind the Earth–Moon pair called the Lagrangian point L2. In this location it is removed from the thermal emissions of the Earth and is capable of much more sensitive measurements of the CMBR. The results have proved the worth of the technical challenges of the mission.

Figure 7.5 shows a comparison of the level of detail of COBE and WMAP.

The fine detail structure predicted by the Big Bang Theory is indeed visible in the WMAP picture of the CMBR of the Universe. The bright spots of yellow and red show the original locations of the dark matter clumps and how they retained the energy-matter plasma in the early stages of the Universe. WMAP has been instrumental in confirming the Big Bang Theory predictions in addition to many other significant successes.

For Interest	'Big Bang'

The term 'Big Bang' is attributed to astronomer Fred Hoyle who disagreed with the idea of a continually expanding Universe. He felt that the Universe was in a steady state and was always like this and referred to the new expansion theory as some sort of big bang. The term has endured ever since.

Consolidation Questions

1   a) Use the data in Table 7.2 to determine the mid-point temperature for each of the stellar classifications from B to K. For types O & M use the minimum or maximum temperature as shown in the table.

    b) For each of the temperatures above, determine the wavelength of peak intensity in the spectrum emitted by that stellar class.

    c) Estimate the position of each peak wavelength ranging between ultraviolet to infrared in the electromagnetic spectrum.

    d) For wavelengths in the visible spectrum, what colour is associated with the peak intensity of the stellar object?

2   Evidence for a Big Bang start to the Universe may be divided into four types. List each type of evidence and summarise the findings which support a Big Bang Theory.

2 Particles and Waves

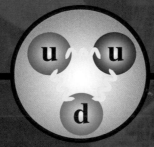

8 Electric fields

When two charged particles interact, they can exert an attractive or repulsive force on each other. The nature of the force depends on the charge, either positive or negative, of each particle. We know from studying magnetism that like poles repel and opposite poles attract. The principle is the same with charged particles: a positive particle will repel another positive particle and a negative particle will repel another negative particle; a positive particle will attract a negative particle and vice versa.

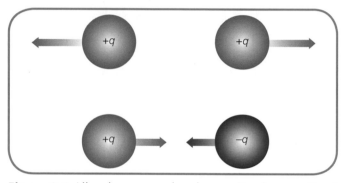

Figure 8.1 Like charges repel and opposite charges attract

Both particles contribute to the force of attraction or repulsion and the magnitude (size) of the force is directly proportional to the size of the charge. The distance between the charges also affects the magnitude of the force – the further away they are from each other, the smaller the force (attraction or repulsion) between them.

It is important to note that these forces occur 'at a distance'. In other words, the two charged particles do not have to touch in order to exert a force on each other. To explain this more clearly we use the concept of an 'electric field'.

Electric fields exist around all charged particles and the field of one particle will be directly affected by the field of every other charged particle around it. We will start this section by considering the field around a single positive charge. Throughout this section we will talk about **point charges**; this means that the charge is of an infinitesimally small size – a point.

You can see from Figure 8.2 that the electric field lines are pointing away from the positive charge. That is

because electric field lines show the direction a positive charge would follow in the electric field. This is further demonstrated in Figure 8.3 which is the equivalent diagram for a negative point charge.

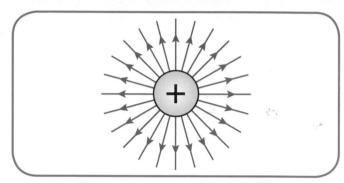

Figure 8.2 Electric field lines around a single positive point charge

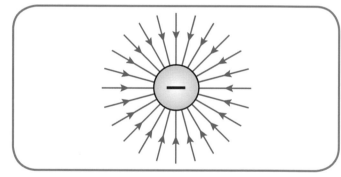

Figure 8.3 Electric field lines around a single negative point charge

These individual electric field lines only show the direction that a positive particle would travel (accelerate) in. They do not show the force on the particle or the strength of the electric field. The diagrams do, however, indicate the relative strength: the closer the lines, the stronger the field. (Note that electric field lines never cross each other.)

Electric field lines have limited use when describing the field around a single point charge, but they become more useful when we introduce an additional point charge. If we have a single positive point charge and a single negative point charge separated by some distance, there will be an electric field between them as shown in Figure 8.4. In this picture, both charges are equal but opposite.

For Interest   Coulomb's Law

The force between two charged particles is given by Coulomb's Law. Coulomb was a French physicist who performed experiments to deduce that the force between particles was proportional to the magnitude of the charge on each particle ($Q_1 \times Q_2$) and inversely proportional to the square of the distance between them ($1/r^2$). The unit of charge, the coulomb, is named after him.

Coulomb's work led to the equation for the force between two particles in a vacuum:

$$F = k\frac{Q_1 Q_2}{r^2}$$

(The constant, k, in this instance is $1/4\pi\varepsilon_0$ where ε_0 is the permittivity of free space.)

There is an obvious similarity between this and Newton's Law of Universal Gravitation. Both describe the force between two bodies and are 'inverse square' laws, but the force on the charges can be repulsive, whereas gravitation is only ever attractive. The inverse square relationship applies to a number of phenomena and is generally the case for phenomena which radiate in all directions.

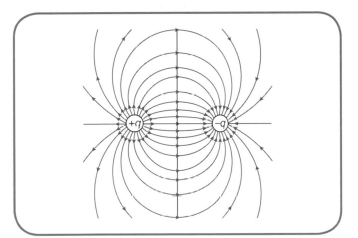

Figure 8.4 Electric field lines between two equal and opposite point charges

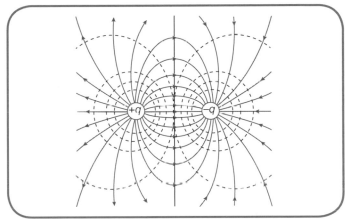

Figure 8.5 Red dashed lines indicate the points at which the electric field strength is the same

It can be useful in this instance to add some extra lines to the diagram. Figure 8.5 is the same case as Figure 8.4, but with the addition of dashed lines which show the position of equal field strength. It is important to remember that the electric field lines show the direction of the field, but the field is everywhere around the charged particle. Each dashed line indicates a position around the particle where the electric field strength is constant.

There will be a similar interaction of electric field lines if both particles have the same charge, as shown in Figure 8.6. This time, instead of leading from the positive to the negative charge as in Figure 8.5, the electric field lines lead away from both charges. This is because, as we mentioned before, the electric field lines describe the direction in which a positively charged particle would travel.

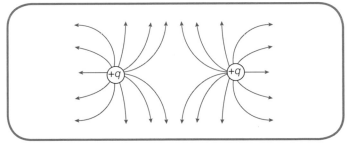

Figure 8.6 The electric field lines show that both of these point charges would repel a positively-charged particle

A further point to note about the diagrams that we have looked at so far is that they are a two-dimensional representation of the electric field. In reality, of course, the electric field 'exists' in three dimensions.

An example of this is shown in Figure 8.7 where the electric field lines surround both of the point charges in

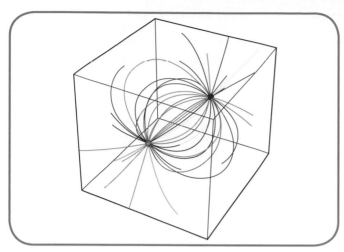

Figure 8.7 A three-dimensional representation of field lines

a three-dimensional representation. Remember that the field lines only show the direction of the field and the relative size (not the value) and that the field surrounds each particle.

Electric field strength

While it is not necessary to calculate the electric field strength for this course, it is useful to understand what affects the field strength. This allows us to have a better understanding of the shape of the electric field lines around a charged particle.

The force on a particle with charge q in an electric field is given by the equation

$$F = qE$$

where E is the electric field strength.

(We can, once again, note the similarity between this and the force of gravity, where the charge q is the equivalent of m and E is the equivalent of g in the equation $W = mg$.) If we rearrange this to make the field strength the subject, the equation can be written as

$$E = \frac{F}{q}$$

This gives us the definition of an electric field – the force per unit charge.

Earlier we saw that $F = k\dfrac{Q_1 Q_2}{r^2}$ where there are two charged particles. Substituting this into $E = F/q$ gives us

$$E = k\frac{Q_1 Q_2}{qr^2}$$

where q is the same as Q_2 if the electric field E is being calculated due to the charge Q_1.

Simplifying this, we reach

$$\text{electric field strength } E = k\frac{Q}{r^2}$$

From this equation we can state that the greater the charge on a particle, the greater the electric field strength around it.

The calculation of electric field strength leads us on to what happens when one of the particles has a greater charge than the other. If this is the case, the field from the more strongly charged particle will have a greater 'sphere of influence', or in other words, it will control more of the space around it, than the weaker particle.

Figure 8.8 shows the situation when the particle on the right has a charge equal to two times that of the other particle. It can be seen that the stronger particle 'controls' more than half of the region between the particles. Compare this with the situation in Figure 8.6 where the control of the region was shared equally.

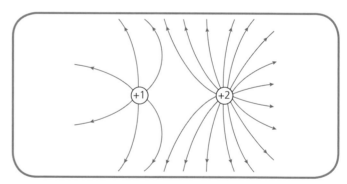

Figure 8.8 Electric field between two unequal charges. The larger charge has a greater effect on the surrounding area

Parallel conducting plates

An electric field surrounds any electric charge. This includes any material, body or shape. One further example of an electric field is that between two parallel charged metal plates (see Figure 8.9). In the middle of the plates, the field is uniform, with straight lines running from the positive plate to the negative plate with equal value. If the charge on either plate is increased, the electric field will increase. At the edges of the plates the electric field lines bend between the plates. This is because the field still has an effect in the areas surrounding the two plates.

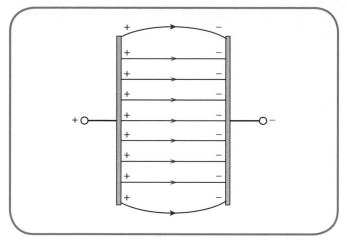

Figure 8.9 Electric field lines between oppositely-charged parallel plates

Activities

This simple experiment will enable you to investigate the shapes of the electric field pattern. Place two charged electrodes in a pool of oil containing a number of small objects such as plant seeds. The plant seeds will move to form the pattern of the electric field around the electrodes. Different charges on the electrodes will cause different patterns and you can use different shapes of electrodes to see how the distribution of charge affects the electric field.

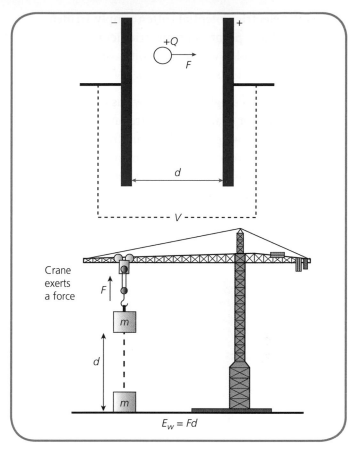

Figure 8.10 As the crane lifts the mass it does work ($E_w = Fd$), in this case to overcome the force due to gravity. Similarly, work must be done to move the particle in the electric field

Movement of charge in an electric field, potential difference and electrical energy

Electrical potential

In the previous section, we investigated the electric field around fixed charges. We now see what happens to particles which are not fixed (they can move) when they are subject to an electric field. To understand this, we must look at the relationship between electric field and the concept of **electric potential energy**.

You should be familiar with the idea of gravitational potential: that is, when you lift something up, the work you do is converted into gravitational potential energy. Moving a mass to a different position within a gravitational field involves work being done.

There is a similar effect with charge in an electric field – if a unit positive charge is moved from a point in an electric field, there must be work done. This work done is the electrical potential energy gained by the single positive charge. This idea is demonstrated in Figure 8.10.

This means that electrical potential energy is a measure of how many joules of work have been done to move the single coulomb of charge to its current position. We may not always want to investigate a unit positive charge, so we need to amend our definition to allow us to investigate the amount of energy required to move any amount of charge.

Written in equation form, this 'electrical potential' would be equivalent to $\dfrac{E_w}{Q}$ where E_w is the electrical potential energy (or work done), measured in joules, and Q is the charge measured in coulombs.

This is the electric potential (often simply written as potential) at a point in an electric field. Potential is measured in volts, where one volt is one joule per coulomb or, as it is usually written, 1 V = 1 J/C.

The potential at a single point allows us to comprehend only so much. For example, if we are on the second level of a building and we pick an apple up from the floor and put it on a table, we will do a certain amount of work, which will be transferred to the apple as gravitational potential. This is not, however, the total gravitational potential energy that the apple 'stores': the apple stores the energy required to take it to the second level and

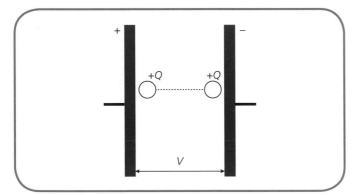

Figure 8.11 On the right, the particle starts with 5 J of kinetic energy. When it reaches the position on the left, it has gained QV more kinetic energy, in addition to the 5 J

the work done to raise it from the floor. When we were thinking about it initially, we were only interested in the energy gained from the floor to the table.

It is very similar when we consider electric potential. We tend to be interested in the amount of energy required to move charge from one point to another, regardless of the electric potential it already holds. This gives rise to the idea of **potential difference**. When we refer to a 'voltage' this is often what we are referring to. Potential difference, measured in volts (V), is the difference in potential between two points.

To describe this idea, we use the equation

$$V = \frac{E_w}{Q}$$

where V is the difference in potential between any two points. In most cases we will define one point as zero and the other point as some positive potential. In a battery, for example, the negative terminal is defined as having zero potential compared to 1.5 V at the positive terminal.

In many cases, we are interested in calculating the work done in the field. In order to do this, we rearrange the above equation to

$$E_w = QV$$

(This is also commonly shown as $W = QV$.)

| For Interest | **The voltaic pile** |

The unit of electrical potential, the volt, is named after Alessandro Volta, an Italian physicist, who is credited with the invention of the battery in 1800. His invention was known as the 'voltaic pile' and was made up of a series of two different metals layered with cardboard soaked in salt water (or acid). This was the culmination of many years of work and was the direct predecessor of modern batteries.

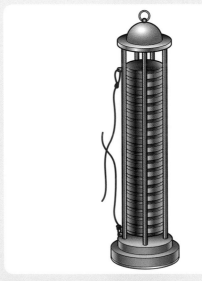

Figure 8.12 The voltaic pile

Movement of charge

Charged particles experience a force in an electric field and this causes them to accelerate. While it is beyond the scope of this course to calculate the acceleration of the charges, if we know how much work has been done in moving that charge it is possible to calculate the velocity of the particle using energy conservation principles.

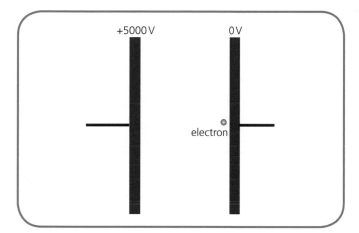

Figure 8.13

To investigate this, we will use the example shown in Figure 8.13. In this case, there are two parallel plates separated by a small distance. The potential difference between them is 5000 V.

We want to investigate what the velocity of the electron would be if an electron is ejected from the 0 V plate, which can be viewed as the 'negative' plate, and moves towards the 5000 V, 'positive' plate. Let us assume the electron is at rest when it is first ejected.

We know that the charge on the electron is 1.6×10^{-19} C and that the potential across the plates is 5000 V. This means that we can use the equation $E_w = QV$ to calculate the work done moving the electron from one plate to the other.

$$E_w = QV$$
$$= 1.6 \times 10^{-19} \times 5000 = 8.0 \times 10^{-16} \text{J}$$

We know that energy must be conserved and, assuming that no energy is lost during the movement of the electron, this means that all of the work done must have been converted into kinetic energy.

The formula for kinetic energy is $E_k = \frac{1}{2}mv^2$ so, as $E_w = E_k$, we can substitute 8.0×10^{-16} J into this equation, along with the mass of an electron which is 9.11×10^{-31} kg.

$$8.0 \times 10^{-16} = 0.5 \times 9.11 \times 10^{-31} \times v^2$$

So $v^2 = 8.0 \times 10^{-16}/4.555 \times 10^{-31} = 1.753 \times 10^{15}$

Taking the square root, this gives us

$$v = 4.19 \times 10^7 \text{m s}^{-1} = 4.2 \times 10^7 \text{m s}^{-1}$$

It should be noted that this result requires two assumptions:

- that gravity has no effect on the system
- that relativistic effects can be ignored.

This use of the principle of conservation of energy could be applied to any particle provided you know its charge and its mass. If the particle had a positive charge, the velocity would be towards the negative plate as opposed to the positive plate, but the calculations involved would be the same.

It is important to remember that in a uniform electric field, the potential is evenly distributed. If there is a potential difference of 100 V between two parallel plates, the potential at the midpoint between them will be 50 V. Using this knowledge we can calculate the velocity of a charged particle anywhere between the plates.

Worked Example 8.1

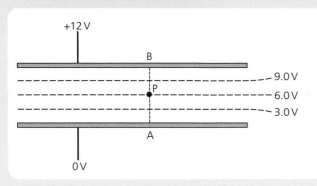

Figure 8.14

Calculate the velocity of a 2×10^{-19} C positive charge at point P if the charge leaves the 12 V plate at rest. The mass of the charge is 5.0×10^{-25} kg.

We know that the particle must move towards the 0 V plate as it is positive. The diagram shows that the midpoint has a potential of 6 V. That means that the potential difference between the 12 V plate and the 0 V plate is 6 V.

We now know the charge and the potential difference so we can calculate the work done in moving the particle.

$$E_w = QV = 2 \times 10^{-19} \times 6 = 12 \times 10^{-19} \text{ J}$$

The next stage is to use the equation for kinetic energy assuming that all energy has been changed into kinetic.

$$12 \times 10^{-19} = \frac{1}{2} mv^2$$

$$v^2 = 12 \times 10^{-19}/(0.5 \times 5.0 \times 10^{-25}) = 4.8 \times 10^6$$

$$v = 2191 \text{ m s}^{-1} = 2200 \text{ m s}^{-1}$$

Questions

1 Draw the electric field lines which would surround the following:

   a) a single negative point charge

   b) two positive point charges separated by distance d

   c) a negative and a positive point charge separated by distance d.

2 If, in the case of question **1c**, the magnitude of the positive charge was double that of the negative charge, how would this affect the field lines?

3 Describe the motion of the particles in Figure 8.15. Explain which of the two particles will move more quickly.

4 An electron is accelerated between the parallel plates shown in Figure 8.16. What kinetic energy is gained by the electron? (Remember the mass of an electron is 9.1×10^{-31} kg and the charge on an electron is -1.6×10^{-19} C.)

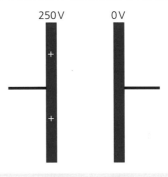

Figure 8.16

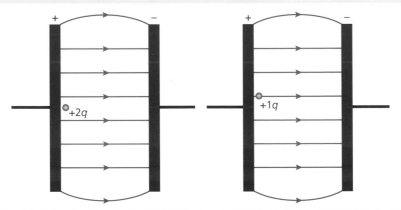

Figure 8.15

5 A proton gains 9.6×10^{-16} J of kinetic energy between two parallel plates. Assuming the electric field is uniform, and the proton originated at the positive plate, what is the potential difference between the plates? (Remember the mass of a proton is 1.67×10^{-27} kg and the charge on a proton is 1.60×10^{-19} C.)

6 The electron in Figure 8.17 will accelerate from rest towards the positive plate.

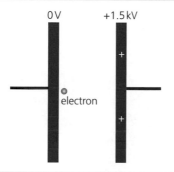

Figure 8.17

Calculate

a) the work done once the electron has reached the plate

b) the velocity at that point.

7

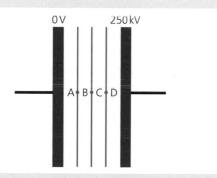

Figure 8.18

a) Calculate the velocity of an electron at points A, B, C and D of Figure 8.18.

b) Are all of your answers realistic? Explain your answer.

8 In a large particle accelerator, lead ions are often accelerated to very high energies. The ions have 82 protons and no electrons and have a mass approximately equal to 208 protons (due to the neutrons). What potential would be required to accelerate a lead ion from rest to two-thirds of the speed of light? State any assumptions you make.

9 Draw a diagram which would indicate the path of a negative ion fired through parallel plates with a 500 kV potential difference between them.

10 What is the definition of a volt?

11 What work would be done in accelerating a two-proton nucleus through a potential difference of 450 V?

12 An 8.5×10^{-27} kg particle is accelerated between the two plates shown in Figure 8.19. The particle moves from the positive plate to the negative plate where it has reached one-third of the speed of light. Relativistic effects can be ignored.

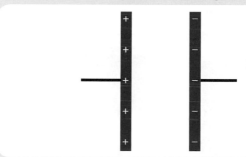

Figure 8.19

a) How much kinetic energy has been gained by the particle?

b) If the potential difference is 88.5 MV, what is the charge on the particle?

c) What is a possible make-up of the particle?

13 Design a system capable of accelerating an electron to half of the speed of light (assuming there are no relativistic effects).

14 A proton enters the parallel plates as shown in Figure 8.20 at high velocity. Suggest a possible path with reasons.

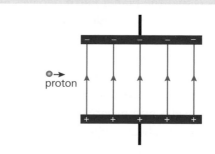

Figure 8.20

Charged particles in a magnetic field

Magnetic fields

Most of us will be familiar with magnets. We find examples of them all around the house – on the fridge, in headphones and in computer hard drives. The history of magnets is a little unclear. There are naturally occurring magnetic rocks which provide the foundation of modern magnets. Some historians believe that in the sixth century BC, or even earlier, there is evidence of the use of magnets for navigation. There is also some mention of iron ore being struck by lightning, but it is difficult to attribute a definitive record of this. Some sources claim it happened in an area called Magnesia.

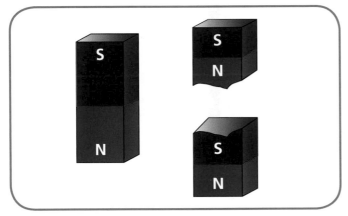

Figure 8.21 A magnet has two poles, one North and one South, but if we split the magnets up, we will have two magnets with two poles each, as individual poles cannot be isolated

Magnetic compasses are an important use of magnets to this day, although advances in technology mean that we are moving increasingly towards more satellite-based navigation devices. The reason why a traditional compass works is that the planet Earth behaves like an enormous magnet. It has two 'poles': the North and the South Pole. It can be useful to compare north and south poles to positive and negative charges in some ways, but what is very interesting is that at no point in history has a single pole been 'isolated'. This means that while we can have positively- and negatively-charged particles on their own, magnets always appear with a north and a south pole.

For Interest **Earth's poles**

There is some misunderstanding about the naming or identification of the Earth's poles. William Gilbert (1544–1603), an English physicist, was the first person to suggest the idea of poles. He described a 'north-seeking pole' and a 'south-seeking pole' and these are what we now describe as north and south poles. As a result, the north pole of a compass will actually be attracted to a south pole. So what we call the North Pole is actually a magnetic south pole and the opposite is true for the geographic South Pole!

Figure 8.22

There are several similarities between poles and charges. If two like poles are placed together, they will repel each other in much the same way as two like charges placed together. If a north pole and a south pole are placed together, they will attract. We call the area which surrounds a magnet the **magnetic field**. Magnetic fields surround permanent magnets and moving charged particles. The effect of a magnetic field diminishes the further away from the magnet you get – this is because the strength of the magnetic field itself is reduced.

Magnetic fields surround magnets and in order to describe them we use magnetic field lines. In this case the field lines show the direction a north pole would follow. (There are obvious similarities here with electric fields.)

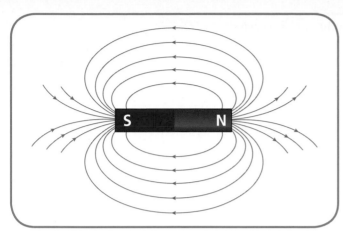

Figure 8.23 The magnetic field around a single bar magnet

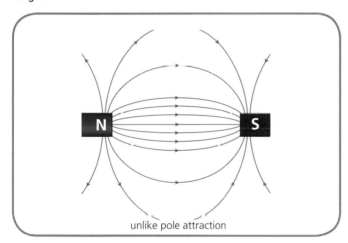

unlike pole attraction

Figure 8.24 The magnetic field between two opposite poles

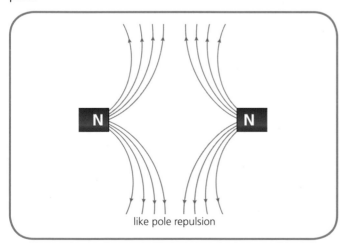

like pole repulsion

Figure 8.25 The magnetic field between two like poles

Figures 8.23, 8.24 and 8.25 show the magnetic field lines in a variety of situations. In each case the lines run from the north pole to the south pole as the lines denote the movement of a theoretical north pole.

Figure 8.23 shows the magnetic field around a single bar magnet, Figure 8.24 shows the magnetic field between two opposite poles and Figure 8.25 shows the magnetic field lines between two like poles. Unless the force caused by the magnetic field is strong enough, a gap between two unlike poles can remain.

Activities

Magnetic effects can be investigated using a very simple experiment.

Place two magnets under a piece of paper.

Pour enough iron filings onto the paper to give a thin coating.

You will already see the effects of the magnets, but you can move the magnets around to investigate the different shapes you can create.

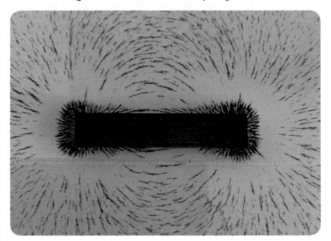

Figure 8.26 Investigating magnetic fields

Research Task

The Earth has a magnetic field therefore it must be a 'magnet'.

Find out what is important about the Earth's magnetic field. Does it affect our everyday lives? What would happen if there was no magnetic field?

As is discussed on page 160, current involves the movement of charged particles. Before we investigate the relationship between current and magnetic fields, however, we must discuss some important points. These points are detailed on the next page.

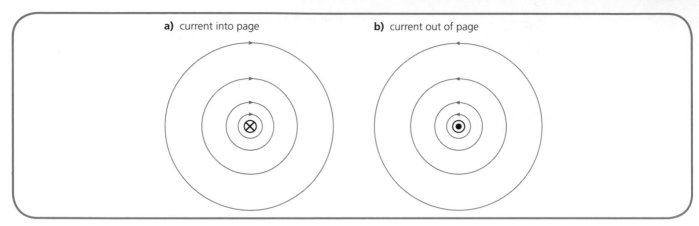

a) current into page

b) current out of page

Figure 8.27 The 'x' shows that the current is going into the page, while the dot represents current coming out of the page

- When describing a situation where the direction of current, magnetic field or other variables is 'into' or 'out of' the page, we use the convention laid out in Figure 8.27. This is necessary as we will discuss phenomena which occur in three dimensions. The direction of the current, magnetic field and force are all interrelated, but occur on different planes.

- The direction of current has been defined as the movement of particles with a positive charge. This is the convention in many textbooks. Unfortunately current in wires is usually due to the movement of electrons which are negatively charged. This means that the flow of negative particles is in the opposite direction to the 'flow' of current.

- When a current is produced in a wire, the electrons are moving through (or vibrating in) the wire. This movement of charge causes a magnetic field to be created around the wire and this can be seen by placing small compasses round the wire. The direction of the current will determine how the compasses respond.

- We can define the direction of a magnetic field around a current-carrying wire using the method shown in Figure 8.28. If you let the thumb of your right hand point in the direction of the current and then wrap your fingers in a fist, the direction in which your fingertips are pointing round the wire is the direction of the magnetic field. This is sometimes known as the right-hand-grip rule.

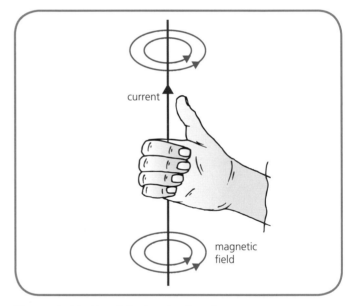

current

magnetic field

Figure 8.28 The right-hand-grip rule

For Interest	Oersted

Hans Christian Oersted (1770–1851), a Danish physicist, is credited with discovering the relationship between current and magnetic field. It is said that his breakthrough came after he observed the movement of a compass needle during a thunderstorm. This led him to experiment with a current passing through a wire placed close to a compass. The compass moved depending on the current, as we can still show for ourselves.

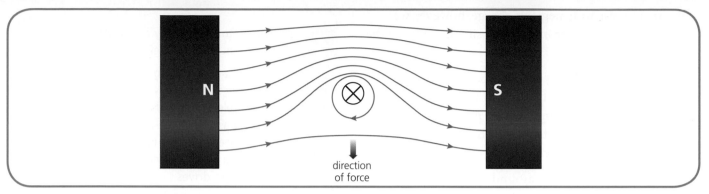

Figure 8.29 The magnetic field of the magnets interacts with the magnetic field of the current-carrying wire. This produces a force as shown, which we call the electromagnetic force

Current-carrying wires in magnetic fields

If a current-carrying wire is placed in a magnetic field, the field created by the current or movement of charge will interact with the magnetic field. This induces a force on the charged particles in the wire (see Figure 8.29).

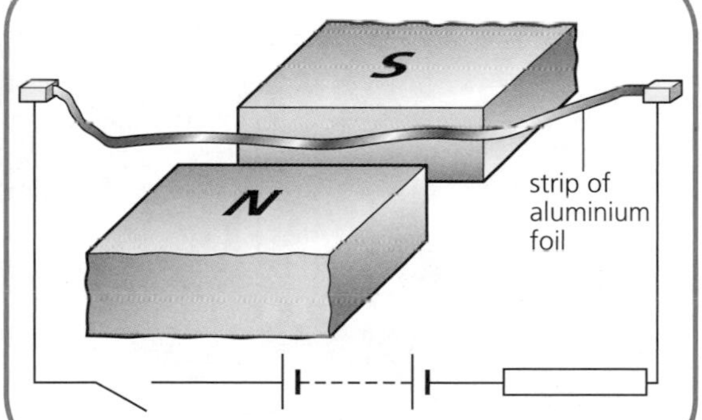

Figure 8.30

Connect a piece of foil to a power supply. Suspend the foil between the poles of a permanent magnet. Observe what happens to the foil when the power is switched on and a current is running through the foil.

What happens when you change the direction of the current? What happens if you switch the poles of the magnet?

(Note: be very careful when performing this experiment; the foil is essentially 'shorting' the supply.)

This experiment leads us to another simple rule – again involving your hands – which helps us to explain the direction of force which moving charge will experience.

As we have seen, a current-carrying wire is seen to move when placed within a magnetic field. It is the moving charged particles within the wire which experience the force. It is the force on each charged particle in the field which moves the foil. This force acts on moving particles, both positive and negative.

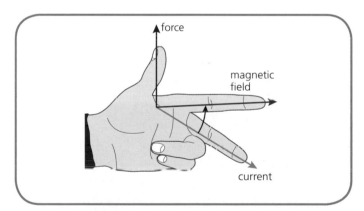

Figure 8.31 The left-hand rule, invented by Sir John Ambrose Fleming, the English electrical engineer and physicist

Positively-charged particles moving in a magnetic field follow Fleming's left-hand rule, as shown in Figure 8.31.

- The thumb indicates the direction of the force on the particle.

- The first finger indicates the direction of the magnetic field.

- The second finger is the direction in which the positive charge is moving.

95

If we are investigating negative charge, however, we point the second finger in the opposite direction to the charge movement.

The force experienced by charged particles in a magnetic field allows us to 'control' positive or negative charges which are moving through magnetic fields. The polarity of electromagnets can be changed by reversing the direction of the current.

By placing electromagnets around an area, as shown in Figure 8.32, and injecting an electron into the fields, we can manipulate the electron by changing the magnetic fields from the electromagnets. Older cathode ray televisions used electromagnets to change the force on electrons to create the television picture, and many oscilloscopes still use this method. It is also the basis for particle accelerators (as further discussed on page 99).

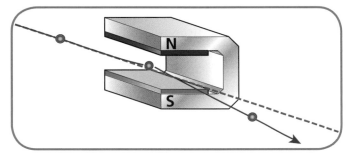

Figure 8.32 The direction of electrons can be controlled by firing them through a magnetic field

Figure 8.33 The curved path of a beam of electrons in a magnetic field. The curve, or deflection, is controlled by the magnetic field

When a beam of charged particles is fired through a magnetic field, it will follow a curved path. The force exerted by the magnetic field will cause a force perpendicular to the direction of travel as we have seen. So this force will cause an acceleration, and thus a velocity, in the direction of the force. This must be combined with the velocity of the particle in a vector calculation to give a path similar to that of a projectile.

Activities

Electric motors are a way of transforming electrical energy into useful kinetic energy.

Electric motors have long been used in cars to start the engine and now we are entering the age of the electric car, they are increasingly being used as the single power source! They have also been used for many years in large systems such as trains and in tiny ones like computer hard drives.

It is possible to build a simple motor in the laboratory with a battery, some wire (some will need to be coated with plastic and some will need to be plain) and a large magnet (or two smaller ones).

Once you have constructed your motor, consider the following questions:

1 Where else might you find a motor?

2 How can a motor be improved over the simple version?

3 Can you control the speed? This can be investigated using different numbers of coils of wire, a variety of batteries or different magnets.

4 Do motors use permanent magnets?

Figure 8.34 Coiled copper wire in an electric motor

Research Task

The force caused by the movement of charged particles inside a magnetic field is widely used today. Most of the energy you use is likely to come from a generator – be it at a wind farm, a more traditional fuel-burning plant or even a nuclear power station. Find out how a generator works. How does this relate to a motor? How are modern generators being made more efficient? Are there different types of generator?

Questions

15 Label the poles correctly in Figure 8.35.

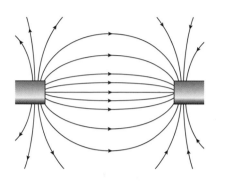

Figure 8.35

16 Sketch the magnetic field lines in the following situations:

a) a single bar magnet

b) two south poles facing each other

c) two opposing poles facing each other.

17 Design an experiment to demonstrate that charges in a wire produce a magnetic field around the wire.

18 In which direction is the current flowing in Figure 8.36, from A to B or from B to A?

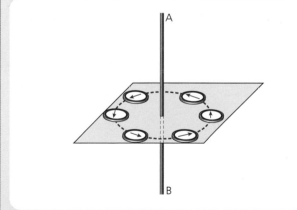

Figure 8.36

19 Indicate the direction of current for each of the examples given below. (Note: ⊗ indicates magnetic field going into the page; ⊙ indicates magnetic field coming out of the page.)

a)

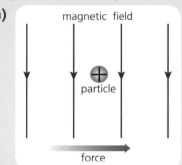

Figure 8.37

b)

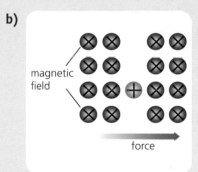

Figure 8.38

c)

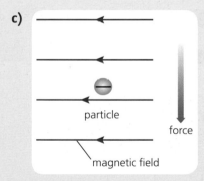

Figure 8.39

d)

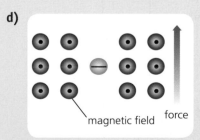

Figure 8.40

20 In each of the cases illustrated below, in what direction is the force on the particle?

a)

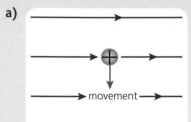

Figure 8.41

b)

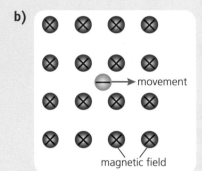

Figure 8.42

c)

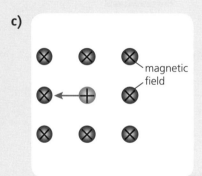

Figure 8.43

d)

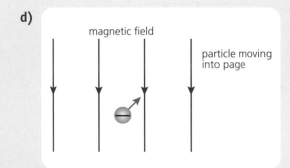

Figure 8.44

21 In each of the cases illustrated below, in what direction is the magnetic field?

a)

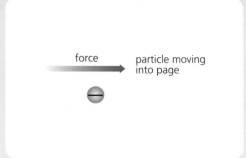

Figure 8.45

b)

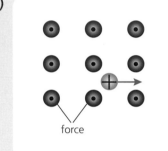

Figure 8.46

c)

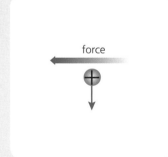

Figure 8.47

d)

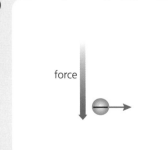

Figure 8.48

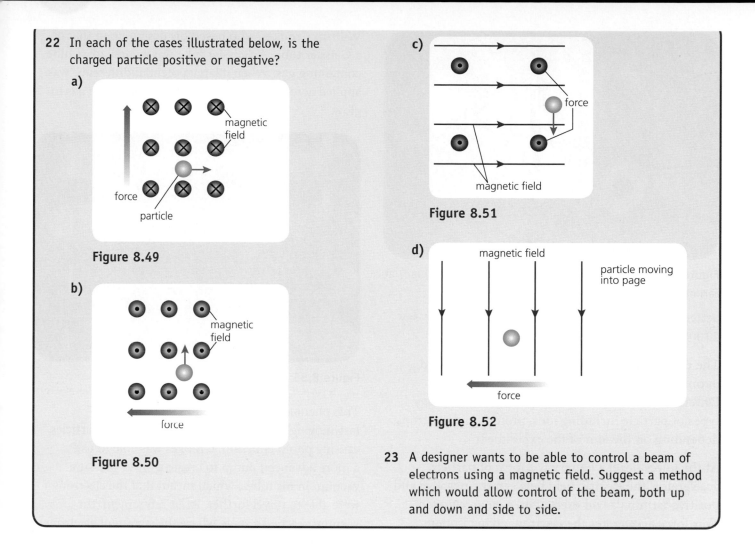

22 In each of the cases illustrated below, is the charged particle positive or negative?

a)

Figure 8.49

b)

Figure 8.50

c)

Figure 8.51

d)

Figure 8.52

23 A designer wants to be able to control a beam of electrons using a magnetic field. Suggest a method which would allow control of the beam, both up and down and side to side.

Particle accelerators

Particle accelerators have a wide range of scientific, industrial and medical applications. Some are enormous like the Large Hadron Collider at CERN but others are much smaller and are fitted onto workstations or small laboratories.

A particle accelerator is a machine which is used to make particles go faster – often with the intention of causing collisions with other particles. The major difficulty is how to accelerate particles to the speeds which are required for certain experiments.

When discussing particle accelerators it is usual to discuss the energy that can be transferred to the particle. The energy is not usually stated in joules, as this is too large a unit for these purposes. Instead the unit **electron volt** or eV is used: $1\,\text{eV} = 1.60 \times 10^{-19}\,\text{J}$. The energies associated with particle accelerators are usually many thousands or millions of electron volts. The unit gives us a

Figure 8.53 Inside the proton linear accelerator (linac) at Fermilab in Illinois, USA

better indication of energy transferred when we say 40 000 eV going to 60 000 eV.

The energy transferred to the particle will depend upon the properties of the particle in question. Different accelerators are set up to accelerate many types of particle including ions, protons and electrons, depending on the aim of the experiment.

At the lowest level a battery is a form of particle accelerator: the electric field between the negative and positive terminals will exert a force on an electron. This force accelerates the electron, giving it more energy. The purpose of a battery is to power a circuit while more sophisticated particle accelerators are used to generate radiation or radioactive samples, or to investigate sub-atomic particles.

Early accelerators

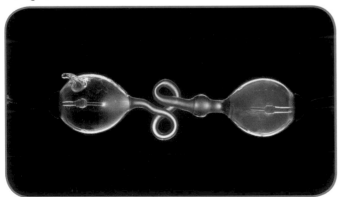

Figure 8.54 Geissler tubes

One of the first devices which was found to accelerate particles was the Crookes Cathode Ray Tube. This device was designed and built in around 1870 by

William Crookes, who based his design upon that of a Geissler tube. A Geissler tube was a small glass tube containing gas. When a large potential difference was applied across the tube, the current caused the gas to glow.

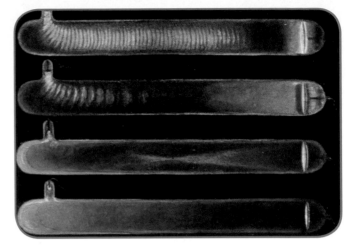

Figure 8.55 Crookes' tubes

This phenomenon occurred as a result of the fast-moving electrons colliding with the gas particles causing photo emission. Crookes was able to use a more advanced pump to create a lower pressure vacuum in his tubes, which meant that the electrons were able to travel further. After refinement, the vacuum reached a stage where the stream of electrons – the cathode ray – would travel the length of the tube, striking the far end and causing fluorescence.

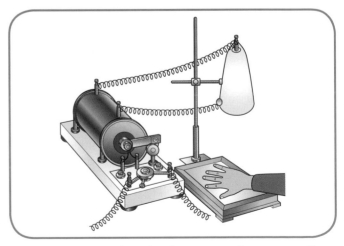

Figure 8.56 A drawing showing Röntgen's apparatus for X-raying his hand

In 1895 Wilhelm Röntgen was studying Crookes' tubes when he noticed something peculiar. When a tube was in use, photographic plates nearby would become clouded. Röntgen realised that this was due to emissions from the tube – emissions which later became known as X-rays. It was for this work that Röntgen was awarded the Nobel Prize in 1901. An adapted version of the Crookes tube is still one of the most widely used methods of X-ray production but the tubes have been altered to focus on the production of X-rays rather than visible light.

At around the same time, J.J. Thomson, a physicist at Manchester University, was also working on cathode ray tubes. He realised he could change the direction of the cathode ray using a magnet (as shown in Figure 8.33). From this work he deduced that there must be a negatively-charged particle which was much smaller than previously considered and that it must form part of the atom. He called it the 'corpuscle' but we now know it as the electron. Thomson was credited with the discovery of the electron and received the Nobel Prize in 1906. It could be argued that this was the first time that a particle accelerator was used in the discovery of a sub-atomic particle.

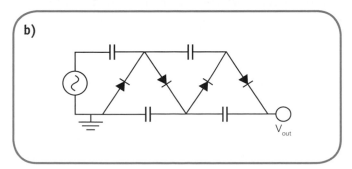

Figure 8.57 a) A Cockcroft-Walton generator and b) Circuit diagram of the generator

In the 1930s, two physicists were conducting experiments with a long evacuated tube. John Cockroft and Ernest Walton built a special generator which produced an 800 kV DC potential from a 200 kV AC supply. This potential was used to accelerate protons towards a lithium target. When the collision occurred it caused the first nuclear transmutation for which they were later awarded a Nobel Prize.

Linear accelerators

Up to this point, accelerators were based on vacuum tubes, with the stream of particles in a straight line. This is how **linear accelerators**, or 'linacs' as they are widely known, work. Modern linacs are no longer made of glass and are carefully prepared vacuum tubes stretching as far as two miles. To understand how these work we can compare them with the analogy of a battery. If we place multiple batteries in series, the total potential difference across them is the sum of each potential difference. The total energy supplied to the electrons will be far greater than that from a single battery.

This idea – the use of a series of individual accelerations – was first used in a linac in the 1920s. In such a linac, particles travel through a succession of charged cylindrical conductors. These 'drift tubes' are placed at different spacings to allow for the rapidly increasing velocity of the particles. The crucial element compared to previous accelerators is the changing field.

The gap between the tubes is the point at which kinetic energy is increased – not inside the drift tubes. Once the particles are almost at the speed of light, the spacing can remain constant. As energy is added to the particles, their mass increases.

If we are accelerating electrons, we must ensure that the next drift tube (relative to the one the particle is currently in) is always positive in comparison. This means that there is an electric field which will exert an attractive force on the electron.

Magnets are placed around the accelerator tube at regular intervals to 'repel' the stream of particles from all sides in order to make it a tightly focused beam. Linacs can be used to reach energies of as much as 50 GeV but this requires a very large distance.

One major problem with linacs is that they are straight. This shape allows relatively simple acceleration and

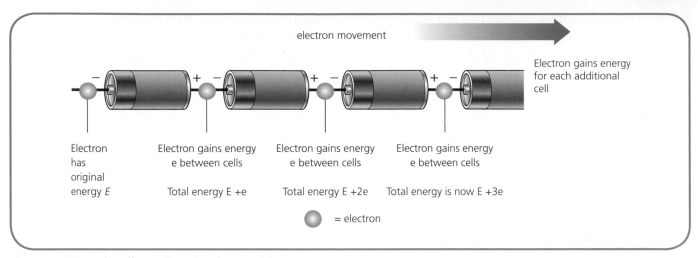

Figure 8.58 Each cell supplies the charge with more energy

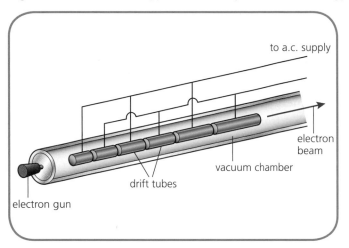

Figure 8.59 An electron linac

Figure 8.60 The first cyclotron, built in 1930

avoids issues of synchrotron radiation which we will discuss in the next section. Because they are straight, however, it means that in order to accelerate particles to very high energy states, the tunnel and the accelerator in it must be incredibly long. It also means that if any particles do not hit their target, they cannot be reused. To counter this some particle accelerators are now circular.

Linacs are still crucial, however, as they are often the first stage when using synchrotrons. They give the particles their initial velocity and the synchrotrons accelerate them to greater velocities.

Cyclotrons

Around 1930, the first **cyclotron**, or circular particle accelerator, was built (Figure 8.60). Designed by Ernest Lawrence, the first prototype was small enough to fit into the palm of a hand, but was only capable of relatively small energies. The system worked very well,

however, and larger and larger versions were made – up to around 18 m radius – all working on the same premise as the original.

Figure 8.61 shows the operation of a cyclotron which is accelerating protons. A cyclotron has an emitter and two 'dees' (metal boxes which are named after their shape) in a vacuum. The dees are supplied with a high-frequency alternating supply, so there is a very large potential difference across the gap between them. The dees are placed inside a strong magnetic field (as shown in Figure 8.61).

Protons are emitted from the centre of the device with some velocity. They follow a curved path due to the magnetic field, as we saw in the previous section, and will accelerate towards the first dee. The curved path continues at constant speed inside the dee until it reaches the gap. It is in the gap that the proton is

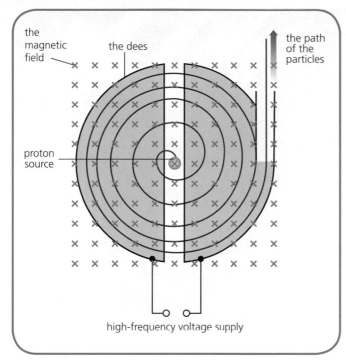

Figure 8.61 The dees of a proton cyclotron

accelerated, not in the dee. The alternating supply frequency is matched to the movement of the particle, so the polarity reverses at the right time to cause the acceleration of the particle across the gap. This process repeats as the particle gains more energy every time it passes the gap. As a result, it follows a wider semicircle each time, before finally exiting the accelerator as a high-energy beam which can be used for collisions.

Cyclotrons have problems, however, related to the production of a constant magnetic field over a large area. Due to relativistic effects (when approaching high velocities), there are problems adapting the changing magnetic field to ensure the correct circular motion.

This problem is dealt with in the **synchrocyclotron**, a slightly adapted version of the accelerator. This device alters the frequency of the supply to match the relativistic mass of the particle. Despite this, such accelerators are still only able to produce around 500 MeV.

Synchrotrons

In order to get much higher energies, particles must be accelerated over a much longer distance. A **synchrotron** has a constant radius no matter what the energy of the particle is – the radius is defined by the shape of the accelerator when it is built. The frequency of the supply and the magnetic field are both altered constantly in

order to maintain the path of the particles as their mass and energy both increase. The deflection caused by the magnetic field makes the path curved, while the frequency of the electric field change increases the energy. A synchrotron is a very complex machine.

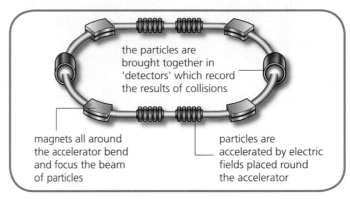

Figure 8.62 Diagram of a synchroton

Synchrotron particle accelerators are currently capable of the highest energies, up to 7 TeV (tera electron volts), but they do suffer from synchrotron radiation. This radiation is emitted and causes a loss of energy as the particle is radially accelerated. There are various methods employed to reduce the effects of synchrotron radiation where it is unwanted. In many experiments, however, it can be used as a reliable source of specific types of electromagnetic radiation. Synchrotrons can be used to generate a specific frequency of X-rays, for example. When used for collision experiments, their closed circle is particularly useful as particles that do not hit their target continue round the circle and another attempt can be made to collide them.

Collision and detection

Magnets are not only used to make particles follow a specific path but also to ensure that collisions occur. Placing magnets around the particle beam focuses it into a slim stream which can be more exactly controlled. This control allows scientists to carry out experiments where two particles can be collided – or smashed together.

There are two methods of doing this: the target can be fixed or it can be another moving beam of particles. The advantage of using two beams in opposite directions is an increase in energy of the collision.

Once a collision has occurred, the particles can split into smaller parts. Detectors surround the point at which the collisions occur. The detectors use magnetic

Figure 8.63 A particle detector at the Large Hadron Collider

Figure 8.64 Particle tracks from lead ion collisions

and electric fields to investigate the results of the collision. The detectors are not set up to detect the particles directly but instead to follow the trace left behind as they ionise the area around them. The paths followed by the ionisation of the detectors give details of the masses and the charges of the particles.

Large-scale particle accelerators

Since the first cyclotron was developed in the 1930s, many accelerators have been created with the sole

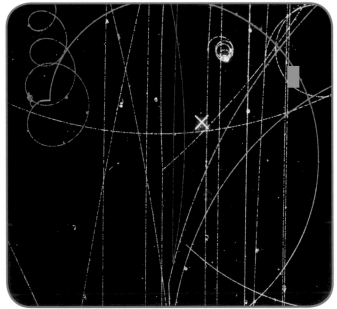

Figure 8.65 Tracks from a bubble chamber particle detector

purpose of scientific research. These have been of all of the various types. In 1962 construction began on the SLAC linear accelerator in America which remains the longest linear accelerator at about 2 miles long. It was completed in 1966 and later that decade was used to provide the first experimental evidence for quarks.

Figure 8.66 The linear accelerator at Stanford, USA

Figure 8.67 Aerial view of the LHC at CERN, Switzerland

The Tevatron in America was another synchrotron, responsible for the discovery of the top quark. The maximum energy of the Tevatron was around 1 TeV and until the Large Hadron Collider was built, it was the highest energy particle accelerator in the world. It was closed in 2011 leaving only one accelerator capable of tera electron volt energies.

The Large Hadron Collider, or LHC, is the world's highest energy particle accelerator. It is part of the CERN complex in Switzerland and, due to its 27 km circumference, part of it is actually in France. For certain experiments particles are accelerated in other accelerators before entering the LHC for further acceleration.

When operating at full capacity, it will be capable of accelerating protons at 7 TeV in both directions, allowing a maximum energy of 14 TeV for collisions. One of the main aims of the LHC project was to search for the **Higgs boson** which was successfully discovered in 2012 (it is an important part of the standard model which may indicate why particles have mass), but the accelerator is part of a larger complex of accelerators and is used for a range of experiments.

Research Task

In this section we have concentrated on the uses of particle accelerators in particle physics. What other uses are there for particle accelerators? Are they always on a large scale? Find out where they have been put to use – some of the results may surprise you!

Questions

24 Why does it take more energy to accelerate a proton than an electron?

25 When designing a cyclotron particle accelerator, an engineer is asked to ensure that the acceleration of the particle remains constant.

 a) What is difficult about this request?

 b) Is there a point at which it will become impossible?

 c) What type of accelerator would you suggest they build instead? (Give reasons for your answer.)

26 Why are the drift tubes in a linac often different sizes?

27 Where in a cyclotron does the acceleration of the particle occur?

28 Synchrotrons often use linear accelerators as the initial stage. What reasons can you think of for this?

29 Explain the following terms with reference to particle accelerators:

 a) acceleration

 b) deflection

 c) collision.

30 It is not normally possible to 'see' a sub-atomic particle. Instead, we tend to use a method of identification which relies on the trace of the particle. Explain how this happens and the methods used to identify the particles.

Consolidation Questions

1 A company has designed an electrostatic spray machine which is used for spray painting vehicles. The paint droplets are charged to 5.6×10^{-16} C inside the machine so that they will be attracted to a negatively charged vehicle. The paint droplets have a mass of 1.1×10^{-9} kg.

As shown in Figure 8.68, the droplets are accelerated from rest on plate X. They accelerate towards plate Y which has holes in it. When they reach plate Y, the droplets are moving at $0.3\,\mathrm{m\,s^{-1}}$.

a) Assuming the effects of air friction are negligible, determine the voltage between plates X and Y for the system to operate correctly.

b) The machine manual states that the electrostatic sprayer ensures a wide spread of paint which will be coated evenly over the vehicle. It also states that less paint will be required. Using your knowledge of physics, comment on these claims.

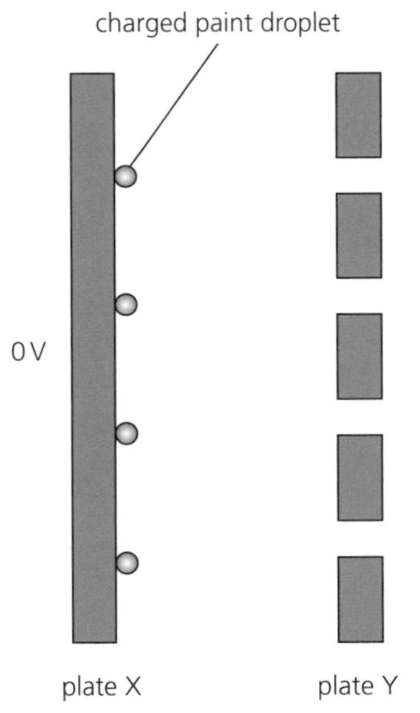

2 A cathode ray tube, used to accelerate electrons towards a screen to create an image, is shown in Figure 8.69 below. The electrons start from rest at the cathode and then accelerate towards the anode.

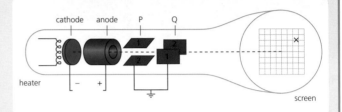

Figure 8.69

The potential difference between the anode and the cathode is 1.5 kV.

a) Calculate:
  i) the kinetic energy of the electron as it reaches the anode
  ii) the velocity of the electron at the anode.

b) P and Q are electromagnets that are used to control the electron beam. In each case, which plate (1 or 2) should be the north pole to cause the beam to divert to point X (at the top right of the grid)?

charged paint droplet

0 V

plate X

plate Y

Figure 8.68

9 The Standard Model

Orders of magnitude

The study of physics involves measurement of the features and phenomena we observe around us. Chronologically, these science measurements started on the same physical scale as ourselves. As humans, we used units which reflected things we could measure easily based on the typical size of an adult (such as inch, feet, cup). As technology has progressed, we are now able to measure quantities on scales vastly different than our own macro (reasonably large) world.

As we progress to measuring the very large or very small it becomes apparent that the familiar rules we employ on our scale do not translate completely to these scales. This chapter deals with measurements of the very small scale (mass and dimension) and the effects observed.

Scientific notation

In order to make sense of and easily manipulate very big or very small numbers we make use of scientific notation. It is important to be able to read and write this notation correctly and the following examples describe its use.

The convention or rules of scientific notation

The speed of light in a vacuum is $2.998 \times 10^8 \, \text{m s}^{-1}$.

The number or **mantissa** (2.998) is always quoted with one, non-zero digit before the decimal place. The number of significant figures in the mantissa reflects the accuracy of the measurement and is limited by the experimental uncertainty.

The multiplier or **exponent** (10^8) is always quoted with an integer power. This is used to scale the number

Distances	$10^{-3} \sim 10^2 \, \text{m}$	distances we can measure comfortably without additional technology
Times	$10^0 \sim 10^2 \, \text{s}$	time intervals we can measure without additional technology
Masses	$10^1 \sim 10^2 \, \text{kg}$	typical human mass range

Table 9.1 Orders of magnitude on the human scale

Distances	$10^{26} \, \text{m}$	distance to furthest known celestial objects
Times	10^{10} years; 10^{17} seconds	time since the Big Bang occurred
Masses	$10^{32} \, \text{kg}$ $10^{50} \, \text{kg} \sim 10^{60} \, \text{kg}$	hyper-star R136a1 (most massive known star) estimated mass of Universe

Table 9.2 Orders of magnitude on a universal scale

Distances	$10^{-10} \, \text{m}$ $10^{-14} \, \text{m}$ $10^{-15} \, \text{m}$ $10^{-18} \, \text{m}$	typical atom diameter typical nucleus diameter proton/neutron diameter electron diameter
Times	$10^{-22} \, \text{s}$	time for photon to cross nucleus
Masses	$10^{-27} \, \text{kg}$ $10^{-31} \, \text{kg}$	proton/neutron mass electron mass

Table 9.3 Orders of magnitude on a sub-atomic scale

up or down to the actual size based on the unit of measurement. In this case the exponent is 100 000 000 – that is 1 followed by eight zeroes. This number may also be called the **order of magnitude** of the quantity. Measurements with the same unit and exponent are said to be of the same order of magnitude.

The speed of light in a vacuum may therefore be written as 299 800 000 m s^{-1} which is cumbersome and may imply a greater accuracy than is appropriate based on the number of significant figures. Writing the value as 2.998 $\times$ 10^8 m s^{-1} avoids these problems.

The mass of an electron is 9.11 $\times$ 10^{-31} kg.

In this case the mantissa is quoted to three significant figures. The exponent is the inverse of 1 followed by 31 zeroes. This means that the mass of the electron on the kilogram scale is 0.000 000 000 000 000 000 000 000 000 000 911 kg – a very small and difficult number to write!

When dealing with such high speeds and/or such low masses very interesting physical effects are observed.

The observed range of orders of magnitude

Based on the human scale the orders of magnitude shown in Table 9.1 are appropriate.

On the scale of the Universe the orders of magnitude shown in Table 9.2 are appropriate.

On the scale of the sub-atomic the orders of magnitude shown in Table 9.3 are appropriate.

Written in tables like these, the numbers lose a little of their underlying meaning. For example, the astronomical distances we deal with are more than a million, million, million, million times larger than those with which we are familiar. At the other end of the scale, the estimated diameter of an electron is a million, million, million times smaller than the typical human scale.

Many people forget or misunderstand the associated meaning of the exponent. A physicist being interviewed on Radio 4 said that a detector could pick up signals of the order 10^{-8} m but that they needed to detect signals of the order 10^{-16} m. 'So you are halfway there then?' said the interviewer. In response, the physicist is heard to emit a sigh.

These huge ranges of distance, time and mass require very sophisticated experiments and technology to measure precisely and accurately. As the equipment and procedures improved, scientists were able to examine the results in greater and greater detail.

Around the beginning of the twentieth century when scientists started probing the sub-atomic world – measuring smaller and smaller dimensions and the physical properties of the particles listed earlier – a startling series of discoveries were made. It became apparent that the particles of the nucleus (nucleons) were in fact composed of a series of more fundamental particles and that the way these particles interacted could be explained in a more unified way than the multiple, separate theories previously proposed.

The Standard Model of fundamental particles and interactions

Particle models of matter are nothing new. One of the earliest recorded was that of Democritus (ca 400BC) based on a thought experiment where he imagined cutting a piece of matter in half, then in half again and so on. He postulated that at some point it would be impossible to divide the matter any further. He proposed this level of being unable to cut any further as being the atom (a-tom meaning indivisible). The atomic model of matter persisted for around 2500 years or so until the late 1800s/early 1900s when the structure of the atom became apparent.

Through Thomson's work on the electron to Rutherford's nuclear model and Chadwick's identification of separate protons and neutrons, the sub-atomic building blocks of the atom were revealed. At the turn of the twentieth century, the newly discovered nuclear radiations (α, β, γ) were used in a number of innovative experiments to further probe the sub-atomic structure.

One such experiment was to look at the kinetic energy of particles associated with α (alpha) and β (beta) radiations. For α it was found that all particles were emitted with (essentially) the same kinetic energy and the exact amount was determined by the radioactive isotope under study. This was appropriate and confirmed the expectation that a quantum state transition was taking place in the nucleus with the energy difference in the two quantum states being transformed into the kinetic energy of the emitted α particle.

The results for β emission, however, were distinctly different. It was found that the β particles were emitted with a range of energies. A typical frequency distribution is shown in Figure 9.1.

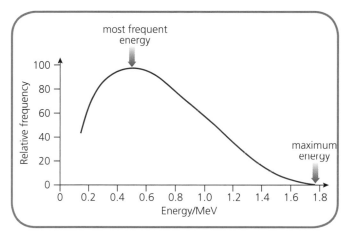

Figure 9.1 A frequency distribution for β particles

The interpretation of these results gives rise to a problem. If the nucleus undergoes the same quantum state change each time the β particle is emitted, then what happens to the 'missing' energy for those particles measured with less than the peak kinetic energy? Why did some particles appear to have more or less energy than others?

In 1930, an Austrian physicist named Pauli proposed that there was another particle being emitted at the same time as the β particle. For this to be true, the second particle had to be of very small mass and with no charge as the particle was not being detected by the experimental methods employed at the time. This proposed particle was subsequently named the **neutrino** by Enrico Fermi, an Italian physicist in the group.

Around the same time as the existence of the neutrino was proposed, another fundamental change to our understanding of the nature of matter came about as a result of the work of an English physicist named Dirac.

Dirac's discovery arose from the fact that the solution of an equation determined by a square root can have two values: positive or negative. In maths that statement is easy and harmless; the square root of 4 is either 2 or −2. In particle physics, however, the consequences are far more wide-reaching. This idea led

to the belief that there are two types of matter: matter and **anti-matter**.

Anti-matter particles have exactly the same characteristics as their matter counterparts with one significant exception; they carry the opposite charge. For example an electron has an anti-matter partner called a **positron**. It has the same mass but the opposite charge. In other words, the positron carries a positive charge. Matter and anti-matter cannot coexist close to each other. If an electron and positron collide then they will annihilate each other and emit energy in the form of radiated photons.

With sufficient energy density, the process can be reversed and a matter/anti-matter particle pair may be produced. This process of interchanging energy and mass is, of course, governed by Einstein's most famous equation, $E = mc^2$.

It turns out that for some complex (quantum mechanical) reasons, the neutrino emitted as part of β decay must be an anti-matter particle – an uncharged, anti-electron neutrino. It was another 25 years before the existence of the neutrino was confirmed experimentally.

(The process described above typifies subsequent advances in sub-atomic particle physics. Theorists examine existing evidence and amend models based on that work, potentially proposing new theories. These new theories may then take many decades to confirm experimentally and the process then repeats itself anew. The search for, and subsequent discovery of, the Higgs boson is perhaps the most well-known example at present, see page 105.)

The structure of the nucleus was now proving more complex than originally thought. A particle much, much lighter than the already very light electron is emitted in concert with that electron. The structure and limit of the nucleus had been further breached. If protons and neutrons somehow decay by emitting these tiny particles, do they themselves comprise still smaller particles? This question was considered by many great physicists at the time. A number of ideas were proposed, discussed and refined.

Subsequent experiments confirmed that this was indeed the case. In the 1960s, American physicist Gell-Mann proposed a model in which the nucleons (protons and neutrons) were composite particles made up from smaller particles called **quarks**. These quarks are thought to be fundamental, which means that they cannot be broken down into further, smaller units.

Gell-Mann called the quarks in protons and neutrons 'Up' and 'Down'. The quark structure of the nucleons is shown in Figure 9.2.

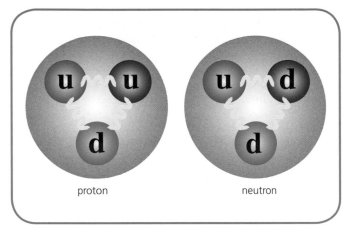

Figure 9.2 Quark structure of protons and neutrons

The proton is a three-quark combination of Up, Up, Down. The neutron is a three-quark combination of Down, Up, Down.

Gell-Mann's work went much further. He proposed three families, or generations, of quarks with increasingly greater mass in each generation (recall that greater mass is equivalent to higher energy). In the current, low-energy conditions of the Universe, only the least massive generation of quarks exist for any significant amount of time.

Each generation of quark has associated with it a much lighter charged particle and neutrino pair. The particles in this pair are known as **leptons** from the Greek word for light. Table 9.4 shows the matter version of the three generations of quarks and leptons along with their electrical charge.

Note that there is another, almost identical, table for the anti-matter quarks and leptons where the charge polarities are swapped. This is given as Table 9.5.

Combinations of quarks form particles called **hadrons** from the Greek word for heavy. There are some simple rules to make the hadrons from quarks:

- Pairs of quarks may be combined to form particles called **mesons**.

- Triplets of quarks may be combined to form particles called **baryons**.

- The electrical charge of the resultant particle must be an integer number.

- Only pairs or triplets of quarks may be used.

From the above, it is clear that mesons must be made from a matter/anti-matter combination. As such, mesons are unstable, short-lived particles. For example, the group of mesons know as pions (π, π^+, π^- in the neutral and charged forms) are made from pairs of matter and anti-matter Up and Down quarks. Their lifetimes are measured between 10^{-16} and 10^{-8} seconds. The psi particle (ψ), consisting of a pair of matter and anti-matter charm quarks, has a mean lifetime of 10^{-20} seconds and proof of its existence led to the 1974 Nobel Prize for two American scientists (Richter and Ting).

It also follows that baryons must be made from either three matter quarks or three anti-matter quarks.

Generation	Name	Symbol	Charge	Name	Symbol	Charge	Name	Symbol	Charge	Name	Symbol	Charge
I	Up	u	$+2/3$	Down	d	$-1/3$	Electron	e	-1	Electron neutrino	υ_e	0
II	Charm	c	$+2/3$	Strange	s	$-1/3$	Muon	μ	-1	Muon neutrino	υ_μ	0
III	Top	t	$+2/3$	Bottom	b	$-1/3$	Tau	τ	-1	Tau neutrino	υ_τ	0

Table 9.4 Matter quarks and leptons

Generation	Name	Symbol	Charge	Name	Symbol	Charge	Name	Symbol	Charge	Name	Symbol	Charge
I	Anti-Up	$\bar{u}$	$-2/3$	Anti-Down	$\bar{d}$	$+1/3$	Positron	$\bar{e}^+$	$+1$	Anti-Electron neutrino	$\bar{\upsilon}_e$	0
II	Anti-Charm	$\bar{c}$	$-2/3$	Anti-Strange	$\bar{s}$	$+1/3$	Anti-Muon	$\bar{\mu}$	$+1$	Anti-Muon neutrino	$\bar{\upsilon}_\mu$	0
III	Anti-Top	$\bar{t}$	$-2/3$	Anti-Bottom	$\bar{b}$	$+1/3$	Anti-Tau	$\bar{\tau}$	$+1$	Anti-Tau neutrino	$\bar{\upsilon}_\tau$	0

Table 9.5 Anti-matter quarks and leptons

These are stable combinations and may be very long lived. Fortunately, we, and the vast majority of the world around us, are made from stable, baryonic matter! Protons are a combination of two Up quarks and a Down quark. Neutrons are a combination of two Down quarks and an Up quark. As a result, the proton carries a unit of $+1$ charge and the neutron is electrically neutral. More importantly, they are very long-lived: free protons are stable and do not decay; a free neutron has a mean lifetime of around 800 seconds. In the bound state of a nucleus, both proton and neutron are equally stable.

How do the particles interact?

The next set of problems or hurdles for scientists to overcome was how do these particles combine, split or affect other particles?

In order to split or attract another piece of matter, some force has to be applied. Think of gravity acting on a stone or a magnet acting on a piece of iron. There are four fundamental, non-contact forces in the Standard Model which may be used to explain how matter interacts. These forces are responsible for attraction or repulsion of matter or charge-carrying matter and are known as:

- the nuclear weak force
- the nuclear strong force
- electromagnetism
- gravitation.

In the Standard Model, the forces which attract or repel particles are explained in terms of a further number of force-mediating particles. That is, the force acting on one object by another is due to a transfer of particles between other, more massive particles.

These force-mediating particles belong to another, larger group of particles called **bosons**. The force-mediating particles are known as **gauge-bosons** and carry momentum and energy between massive particles.

For repulsion, for example, imagine that a massive particle A emits a gauge-boson which impacts and is absorbed by a second massive particle B. This will impart momentum in the impact and cause B to move away from A. In addition, as it emits the gauge-boson, A will recoil away from B in order to comply with the principle of conservation of momentum (as detailed in Chapter 3). When viewed from the outside, this system of particles (A and B) appears to repel. Similarly, if A ejects a boson in the opposite direction, then it will appear to move towards B, in other words, be attracted to B. These interactions are depicted in Figure 9.3.

This appears difficult to understand but this model and the idea of these particles came out of years of experimentation, analysis, further refinement and then more detailed experimentation.

The four forces have dramatically different ranges. The two nuclear forces only operate on the nuclear scale (up to about 10^{-15} m separation) and then they become essentially zero magnitude. The two other forces operate over infinite separation (in the macro-world, the force equation for gravitational or electromagnetic interaction is of the form $F \propto 1/r^2$).

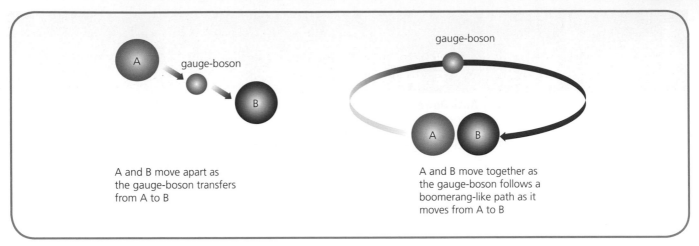

Figure 9.3 Gauge-boson interactions

	Relative strength	Range/m	Gauge-boson
Strong	1	10^{-15}	gluon
Electromagnetic	$\dfrac{1}{137}$	infinite	photon
Weak	10^{-6}	10^{-18}	W^+, W^-, Z bosons
Gravitation	10^{-39}	infinite	graviton (predicted)

Table 9.6 Properties of the gauge-bosons

The strengths, ranges and names of the gauge-bosons are listed in Table 9.6.

Notice how weak the gravitational interaction is compared to all the other forces. It is at least 33 orders of magnitude smaller than any other force. The reasons for this weakness and the existence of the graviton have still to be determined experimentally.

The matter particles (quarks and leptons) belong to a family of particles known as fermions. The force exchange particles belong to a family of particles known as bosons. This classification is based on a quantum property of particles known as spin.

In 2013 a major breakthrough was announced by CERN scientists reporting from the Large Hadron Collider. Forty-nine years after Professor Peter Higgs at Edinburgh University predicted the existence of the Higgs boson, scientists have confirmed that this particle exists with a mass of around 125 GeV – some 133 times more massive than a proton.

The Higgs was proposed as an explanation for the fact that gauge-bosons of the weak force (the W and the Z bosons) have mass, but the gauge-boson (the photon) of the electromagnetic force is massless. This break in symmetry between the weak and the electromagnetic forces is a result of the Higgs field and its associated Higgs boson.

Consolidation Questions

1 How many orders of magnitude are there between the smallest and the largest known

   a) masses

   b) times

   c) lengths?

2 Copy and complete Table 9.7.

3 Sketch the shape of the graph of frequency of beta particle kinetic energy measured for a radioactive decay. (Remember graph sketches need axes labels and units along with any numbers you can include.) What conclusion can be drawn from the shape of this graph?

4 What is meant by a baryon? Make a list of all possible baryons using only the quarks from the first two generations (i.e. Up, Down, Charm and Strange). Use the internet to name your newly discovered matter particles.

5 What is meant by a meson? Make a list of all possible mesons using only the quarks from the first two generations (i.e. Up, Down, Charm and Strange). Use the internet to name your newly discovered matter particles.

Quantity	Value	Scientific notation
electron charge	0.000 000 000 000 000 000 16 C	
speed of light		$3 \times 10^8\,\mathrm{m\,s^{-1}}$
average Earth–Sun distance	150 000 000 000 m	
typical nucleus diameter		$1 \times 10^{-15}\,\mathrm{m}$

Table 9.7

10 Nuclear reactions

Isotope naming

Any element in the periodic table may be identified by its atomic number, defined as the number of protons in the nucleus. This number determines the chemical properties of each atom of the element. The mass number of each element is the total number of nucleons (protons + neutrons) in the nucleus. An element may then be described by the following symbols:

$$^{A}_{Z}X$$

where A = mass number; Z = atomic number; X = element symbol.

For example, $^{4}_{2}He$ describes a helium atom with two protons and two neutrons, in other words four mass units, in the nucleus. In fact, either the atomic number (Z) or the element symbol may be used to establish which element is being described.

Note that it is possible to have different versions of the same chemical element (in other words, with the same atomic number) with different mass numbers. This means that their nuclei consist of the same number of protons but some have different numbers of neutrons. Such nuclei are known as **isotopes**.

Isotopes may have any number of neutrons but must have the same number of protons in the nucleus. It turns out that the most stable isotopes have similar numbers of protons and neutrons. The periodic table shows the most stable (most long-lived) isotope of each element. Many other isotopes are possible for each element. Figure 10.1 shows the known isotopes for each element.

Note that the number of protons is plotted on the x-axis and the number of neutrons is plotted on the y-axis. The stability of isotopes with equal proton and neutron numbers is apparent from the shape of the graph at low atomic numbers. As atomic number increases, this approximate relationship breaks down.

Energy/mass equivalence

In 1905, a young patent clerk from Germany published four scientific papers in some of the world's most prestigious scientific journals. The clerk was Albert Einstein. His published papers included perhaps the most famous equation in the world: $E = mc^2$. Simple and elegant with only three terms, the true meaning of this equation is easy to overlook. In a single equation, Einstein had made the boldest of statements: that energy and mass are equivalent! The c^2 part of the equation is a scaling factor. Einstein claimed that if it were possible to convert 1 kg of mass to energy, then the energy produced would be 9×10^{16} J. Therefore, if we could find a way to convert mass easily to energy, then our demands for domestic and commercial electrical energy could be easily satisfied.

Of course, the issue lies in that final sentence. How might mass be converted to generate energy? Equally, how might energy be converted to generate mass in the form of particles?

Having determined the principle that energy and mass are interchangeable, the key for scientists was how to tame nature and effect these interchanges.

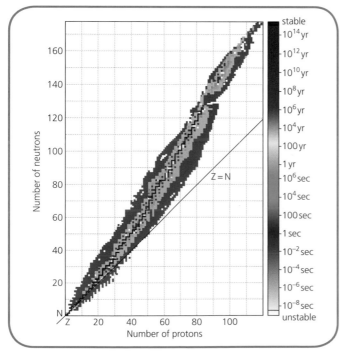

Figure 10.1 Graph of known elements and isotopes

The examination of the nature of nuclear radiations provided the path to determining how this equation might be applied to practical ends.

Binding energy: the 'missing' mass

Detailed experimentation allowed scientists to measure the mass of individual protons and neutrons. When the mass of these nucleons in combination was measured, however, the combined total was found to be less than the simple sum of the individual masses. For example in the $^{12}_{6}$C nucleus there are six protons and six neutrons. This should give a total mass of $(6 \times 1.673 \times 10^{-27}) + (6 \times 1.675 \times 10^{-27})$ $= 20.09 \times 10^{-27}$ kg.

In reality the atomic mass of $^{12}_{6}$C is found to be 19.9×10^{-27} kg!

The 'missing' mass is known as the **mass defect**. The stable combination of protons and neutrons in the nucleus is less massive as the nucleus has lower potential energy than the separate nucleons. At the time of combination, in other words the creation of the atomic nucleus, this energy is radiated away, typically as gamma radiation.

It is theoretically possible to combine any number of protons and neutrons to create a nucleus. Figure 10.2 shows the known combinations of protons and neutrons for each element. Notice that for any vertical line, the atomic number remains constant. In other words, the nuclei on this line are chemically the same element with the same number of protons but varying numbers of neutrons. These are isotopes, as discussed earlier.

The most stable isotope (the one with the greatest mass defect for each atomic number) is shown in red. Other known isotopes are shown in blue.

The binding energy for any isotope of any element is calculated using Einstein's equation $E = mc^2$, where m in this case is the mass defect for that isotope. It is possible to plot a graph of binding energy per nucleon versus mass number for the most stable isotope of each element and such a graph is shown in Figure 10.3.

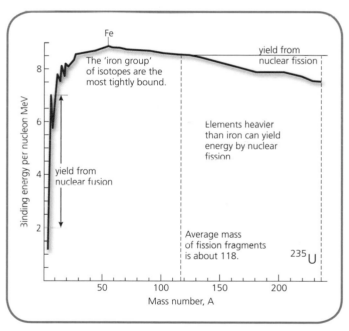

Figure 10.3 Graph of binding energy against mass number

This is a very important graph for nuclear reactions. Note that as mass number increases, the binding energy per nucleon also increases for elements with a mass number less than that of iron (Fe). After this point, the binding energy per nucleon begins to decrease steadily as the mass number becomes larger.

This means that from an energy perspective, it is possible to combine small nuclei to create larger nuclei (nuclear fusion) with a net release of energy and a

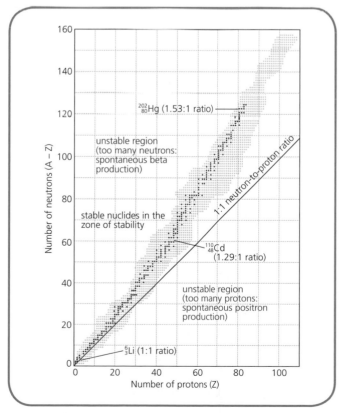

Figure 10.2 Graph of neutrons against protons

correspondingly more stable resultant nucleus. This is true provided that the fusion of the smaller nuclei results in a nucleus whose mass number is less than that of iron. For nuclei larger than that of iron, it is energetically advantageous to split (nuclear fission) to create smaller nuclei whose mass number is at least that of iron.

The two processes, nuclear fusion and nuclear fission, are fundamental to the energy distribution of the Universe and our existence on Earth. The fusion process fuels our star (the Sun) and all other stars in the Universe. The fusion of hydrogen nuclei to create helium with the corresponding release of energy is responsible for virtually all of the energy which reaches us from the Sun. Once all the hydrogen in a star is converted to helium, helium fusion takes over as the main energy-producing process. After helium comes lithium and so on with each increasing atomic number. Stellar processes are incapable of fusing elements heavier than iron until the 'death throes' of the red giant/supernova phase where the excess energy required to create these nuclei results from the extreme conditions within the collapsing star. Once these heavier-than-iron nuclei are created, the subsequent explosion of the star distributes them throughout the Universe. In this way, it is true to say that we are all made from recycled stellar dust. All of the elements on Earth, including those comprising our bodies, were once part of a previous star and formed as it died.

Fission and fusion are possible on Earth. The conditions required to allow nuclear fission are easier for us to recreate than the stellar conditions necessary for nuclear fusion. The following sections look at each of the processes in detail.

Nuclear fission

In early experiments carried out around the beginning of the twentieth century, it became apparent that odd things were happening when nuclear radiations took place. Experiments into the products of nuclear radiation showed that heavy elements (those with masses significantly greater than iron) emitted nuclear radiations and, in certain cases, resulted in the creation of different elements of lower mass than the original element. In effect, large-mass atoms spontaneously break apart into smaller-mass atoms and emit nuclear

radiation in the process. This process is known as **fission**.

Given that the radiation carries energy, this would seem to negate the law of conservation of energy. Detailed analysis of the mass of the atoms before and after fission, however, showed that the total mass before and after was different. This would appear to negate the law of conservation of mass. Of course the answer lies in Einstein's famous equation. If the difference in mass, m, before and after the emission is used in the equation $E = mc^2$, the energy of the emitted radiation is given by E.

For example, $^{236}_{92}U$ decays according to the equation:

$$^{236}_{92}U \rightarrow {}^{144}_{56}Ba + {}^{89}_{36}Kr + 3\,{}^{1}_{0}n$$

Detailed analysis of the atomic mass of the reactants and products in this nuclear reaction gives us the information shown in Table 10.1.

Element	Mass/kg
$^{236}_{92}U$	3.91963×10^{-25}
$^{144}_{56}Ba$	2.38990×10^{-25}
$^{89}_{36}Kr$	1.47651×10^{-25}
$^{1}_{0}n$	1.67493×10^{-27}

Table 10.1 Uranium fission reaction masses

Using the information from the table to analyse the mass before and after fission shows the following:

Mass before: $3.91963 \times 10^{-25}\,kg$

Mass after: $2.38990 \times 10^{-25} + 1.47651 \times 10^{-25} + (3 \times 1.67493 \times 10^{-27})$

$= 3.91666 \times 10^{-25}\,kg$

Mass difference:
$$\begin{array}{r} 3.91963 \times 10^{-25}\,kg \\ -3.91666 \times 10^{-25}\,kg \\ \hline 0.00297 \times 10^{-25}\,kg \end{array}$$

$= 2.97 \times 10^{-28}\,kg$

Energy release per atom: $2.675 \times 10^{-12}\,J$ (as $E = mc^2$)

The mass defect in this reaction realises a large amount of energy with sufficiently large populations of atoms. This reaction is the basis of most nuclear power stations and nuclear weapon technologies.

$^{236}_{92}U$ nuclear fission

Given its inherent instability, the uranium isotope $^{236}_{92}U$ is relatively rare on Earth; however, it is possible to create $^{236}_{92}U$ by adding a neutron to the more stable isotope, $^{235}_{92}U$. The nuclear reaction may be represented by the equation

$$^{235}_{92}U + ^{1}_{0}n \rightarrow ^{236}_{92}U \rightarrow ^{144}_{56}Be + ^{89}_{36}Kr + 3^{1}_{0}n$$

Notice that the products of the reaction include three additional neutrons. If these neutrons can combine with other $^{235}_{92}U$ nuclei, then we may rapidly build a chain reaction where each step in the chain grows by a factor of three: that is three times more neutrons, three times more nuclei 'fissioning' and three times more energy released at each step of the chain. If left unchecked, this chain reaction is capable of releasing large amounts of energy in a very short space of time. In addition to the energy released from the fission of the uranium, the other products of the reaction ($^{144}_{56}Be$ and $^{89}_{36}Kr$) are also unstable and their nuclei decay into further unstable, radioactive products. This reaction was the basis of early nuclear weapons and proved to be a devastatingly fatal technology.

In nuclear power stations, this reaction may be carefully controlled to release energy in a more managed fashion. The nuclear energy generation technology in the UK is based on this reaction but there are several control measures in place to determine how the reaction might proceed:

- Moderator: this is used to determine how quickly the neutrons produced by the reaction travel. By moderating the neutron speed, the probability of collision and inclusion into a $^{235}_{92}U$ is increased. Without the moderator, very few of the neutrons would be slow enough to be captured by a $^{235}_{92}U$ nucleus.

- Control rods: these are used to control how many of the neutrons produced at each step of the chain reaction are capable of reacting with a $^{235}_{92}U$ to produce $^{236}_{92}U$. The control rods absorb most of the neutrons produced and without them, the nuclear reaction would proceed to an uncontrolled chain reaction.

- Cooling system: this is used to transfer the thermal energy produced by each nuclear fission away from

the reactor and into a conventional steam generator/turbine combination to generate electrical energy.

Other safety considerations in the reactor design are also required. The reactor itself is housed in a containment vessel to prevent escape of radioactive material and nuclear radiations.

Nuclear fusion

Energy may also be derived from a **fusion** reaction which forces two low-mass nuclei together to form a more massive nucleus: two hydrogen atoms forming a helium atom, for example. Accurate mass measurements before and after the fusion reaction show a similar mass defect to that in a fission reaction. The total mass of the reactants is greater than the total mass of the products.

For example, $^{3}_{1}H$ may combine according to the equation:

$$^{3}_{1}H + ^{3}_{1}H \rightarrow ^{4}_{2}He + 2^{1}_{0}n$$

Detailed analysis of the atomic mass of the reactants and products in this nuclear reaction gives us the information shown in Table 10.2.

Element	Mass/kg
$^{3}_{1}H$	5.00827×10^{-27}
$^{4}_{2}He$	6.64648×10^{-27}
$^{1}_{0}n$	1.67493×10^{-27}

Table 10.2 Hydrogen fusion reaction masses

Using the information from the table to analyse the mass before and after fusion shows:

Mass before: $10.01654 \times 10^{-27}\,kg$

Mass after: $9.99634 \times 10^{-27}\,kg$

Mass difference:
$$\begin{array}{r} 10.01654 \times 10^{-27}\,kg \\ -9.99634 \times 10^{-27}\,kg \\ \hline 0.02020 \times 10^{-27}\,kg \end{array}$$

$$= 2.020 \times 10^{-29}\,kg$$

Energy release per atom: $1.818 \times 10^{-12}\,J$

The mass difference in this reaction releases a large amount of energy with sufficiently large populations of atoms (remember that a mole contains 6×10^{23} atoms).

$^3_1$H nuclear fusion

With sufficient initial kinetic energy it is possible for two $^3_1$H nuclei to collide and the nuclei to be forced together to create a new, more massive helium nucleus. ($^3_1$H is known as tritium.) In stellar conditions, the huge gravitational field strength creates conditions which bring tritium nuclei together with sufficient energy for fusion to occur.

In order to recreate these conditions on Earth there are very large problems to overcome. Recall that the average kinetic energy of the atoms or molecules of a substance may be represented as the temperature of the substance. To make the conditions where kinetic energy is sufficiently high for fusion to occur is the equivalent of a temperature of the order of 10^7 K. At such temperatures the molecules of a gas will lose electrons to become positively-charged ions. It is not possible for such a plasma to be contained in a normal material vessel as the walls would melt or evaporate on contact with the plasma itself. The high-temperature and high-chemical reactivity of the plasma means that

a novel way of keeping it contained is necessary. This is solved by the use of a strong magnetic field.

In Chapter 8 we discussed the fact that moving charged particles in a magnetic field will be subject to a force perpendicular to their direction of travel. This central force constrains the particles to move in a circle. Making use of this effect, the first designs for a plasma containment vessel were realised in Soviet Russia. Their TOKAMAK reactors were invented in the 1950s by Soviet physicists Igor Tamm and Andrei Sakharov. The name TOKAMAK is an acronym for the Russian phrase which means 'a toroidal chamber with magnetic coils'.

Achieving a working fusion reaction which is both sustainable and realises a net delivery of useful energy has proved problematic. At the time of writing, the world's largest and most advanced tokamak reactor is under construction near Cardarche in the south of France.

As with CERN, this is an international collaboration. Known as ITER (International Thermonuclear Experimental Reactor), it is planned to be the stepping stone from research to commercially viable electrical energy generation producing output power of 500 megawatts. In order to achieve this, the power required to ignite and sustain the plasma is 50 MW! Along with the engineering constraints in building the magnetic field cage to contain the plasma, designing a mechanism to extract energy from the plasma as the nuclear reaction proceeds, and ensuring a sustainable delivery of output power, the enormous amount of power required to achieve fusion makes this an exceptionally difficult project to complete. The ITER is planned to switch on around 2025 with the first commercial power plant, DEMO, to follow thereafter.

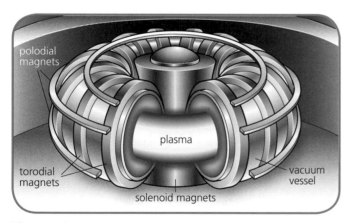

Figure 10.4

Consolidation Questions

1 Use Table 10.3 to estimate the kinetic energy of the alpha particle emitted in the following reaction:

$$^{234}_{90}\text{Th} \rightarrow {}^{230}_{88}\text{Ra} + {}^{4}_{2}\text{He}$$

2 A possible fission of uranium-236 gives rise to strontium-88 and xenon-136 and multiple neutrons.

   a) Write down the equation for the nuclear reaction.

   b) What is the energy released per reaction? Use Table 10.4 to work this out.

Element	Mass/kg
thorium-234	3.88638×10^{-25}
radium-230	3.81986×10^{-25}
helium-4	6.64647×10^{-27}

Table 10.3

3 Sketch the shape of the graph of binding energy per nucleon versus mass number. (Remember graph sketches need axes labels and units along with any numbers you can include.) Label the graph to show regions where:

   a) fusion is energetically possible

   b) fission is energetically possible.

   c) What is the most stable nuclear species (i.e. the one least likely to decay by fusion or fission)?

Element	Mass/kg
uranium-236	3.91974×10^{-25}
strontium-88	1.45971×10^{-25}
xenon-136	2.25679×10^{-25}
${}^{1}_{0}\text{n}$	1.67492×10^{-27}

Table 10.4

(11) Wave properties

The topic of waves is a large one and while the majority of this section will focus on interference, diffraction and refraction, we will start by investigating a few of the basic principles of waves.

Waves are all around us – light waves, radio waves and sound waves. Sound waves are **longitudinal waves**, or **compression waves**, while radio waves, light waves and the rest of the electromagnetic spectrum are **transverse waves**. Figure 11.1 shows the difference between the two types of wave.

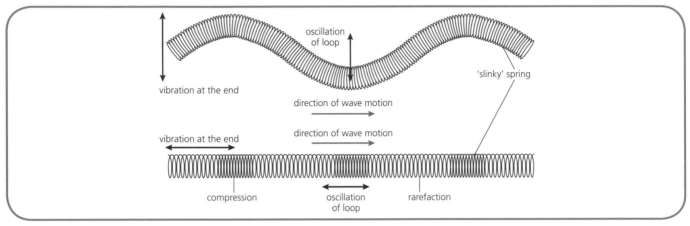

Figure 11.1 In a transverse wave, such as a light wave, the vibrations are perpendicular to the direction of travel. In a longitudinal wave, such as a sound wave, the vibrations travel in the same direction as the wave – this is why it is also known as a compression wave

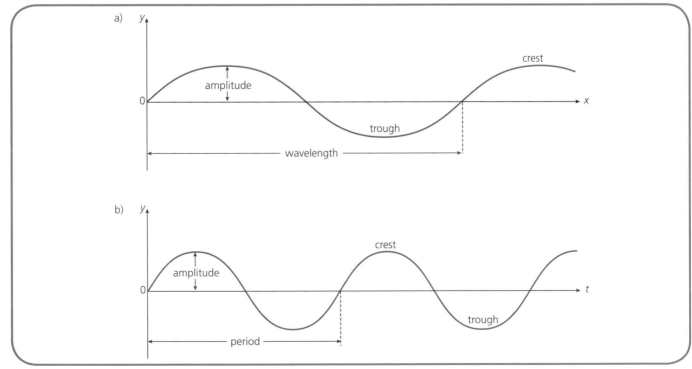

Figure 11.2 a) Displacement against distance; b) displacement against time

We will deal mainly with transverse waves in this section, since light waves allow us to investigate wave behaviour with something we can see.

Some important attributes of waves are shown in Figure 11.2.

- Peaks occur at the top and the bottom of the waveform. A positive peak is often referred to as a **crest** and the negative peak as a **trough**.

- A is the **amplitude** and is defined as the distance from the centre to the peak of the wave.

- T is the **period**, defined as the length of time a single wavelength or cycle of the wave lasts. It is measured in seconds.

- f is the **frequency**, defined as the number of cycles of the wave which pass a point in a single second. It is measured in Hz.

The frequency and period are linked by the equation:

$$T = 1/f$$

Using the equation for the relationship between displacement, velocity and time, $s = vt$, we can compare displacement with wavelength, and time with period.

Using $T = 1/f$ (this is the period or time for one wave) and substituting in $s = vt$, we get $s = v/f$. This is the displacement which occurs in the time for the wave to go from crest to trough and back to crest again. The displacement of one wave is its **wavelength**, λ. This gives.

$$\lambda = v/f \quad \text{or} \quad v = f\lambda \text{ (as it is normally written)}$$

where

- v is the velocity of the wave, measured in m s^{-1}

- λ, the Greek letter lambda, is a measure of the length of the wave (the wavelength), measured in metres.

This relationship is called the **wave equation**.

Worked Example 11.1

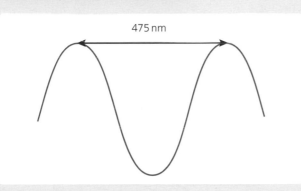

475 nm

Figure 11.3

A light wave of wavelength of 475 nm is travelling in a vacuum.

Calculate

a) the frequency   b) the period.

The first step in this question is to note that it is a light wave in a vacuum, and so it must be travelling at

$3 \times 10^8\,\text{m s}^{-1}$. Using this we can write

$v = 3 \times 10^8\,\text{m s}^{-1}$

$f = ?$

$\lambda = 475\,\text{nm}$

We can see that the wave equation will give us the frequency.

$v = f\lambda$

$3 \times 10^8 = f \times 475 \times 10^{-9}$

$f = 3 \times 10^8/475 \times 10^{-9}$

$f = 6.32 \times 10^{14}\,\text{Hz}$

given $T = 1/f$

$T = 1/6.32 \times 10^{14} = 1.58 \times 10^{-15}\,\text{s}$

Wave properties

All waves have a number of properties in common. They can:

- reflect

- refract

- diffract

- interfere.

The fact that waves reflect is something that we are aware of in our everyday lives: when we look in a mirror or hear an echo, for example. Both of these are examples of reflection. When a wave reflects on a surface, the **angle of incidence** (as shown in Figure 11.4) is always the same as the **angle of reflection**. To show this, we use an imaginary 'normal' line which is perpendicular to the surface of the reflecting medium at the point where the wave hits the medium.

Reflection is the property of waves used in satellite transmission systems where a curved reflector focuses multiple waves at a single point. Figure 11.5 shows how the signals are focused.

As shown in Figure 11.5, the normal is always perpendicular to the surface at the point of incidence, so for each wave the angle of incidence is different. As a result, their reflections meet at a single point. The angle of the curve can be changed to alter the point of focus, depending on what the curved reflector will be used for. For television satellite dishes, a detector is placed at the focus. This receives lots of different signals which are used by the decoding boxes. The same system can be used in reverse to create a transmitter, as shown in Figure 11.6. The detector is replaced with an emitter and the signal is reflected as parallel rays from the reflector.

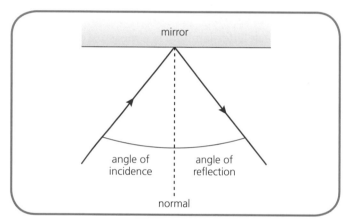

Figure 11.4 How light reflects – the angle of reflection is the same size as the angle of incidence

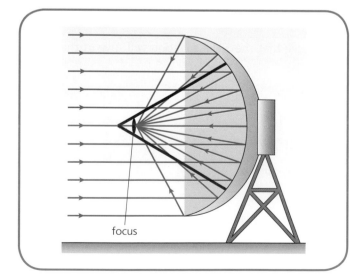

Figure 11.5 A curved reflector reflects signals, such as radio or light, and focuses them at a single point. This strengthens the received signal

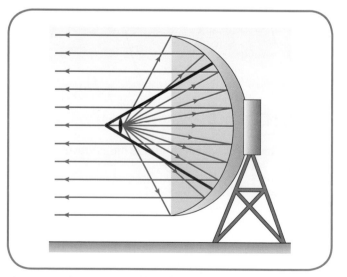

Figure 11.6 A curved reflector is also used for the transmission of signals. The signals from the single emitter hit the curved surface at different angles, causing their reflected waves to be parallel

Figure 11.7 An array of parabolic reflectors

Questions

1 Copy Figure 11.8 and add labels for amplitude and wavelength.

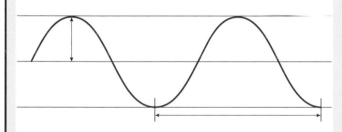

Figure 11.8

2 A 50 Hz wave has a wavelength of 2 m. What is its velocity?

3 What is the frequency of a 2 cm wave travelling at $2 \times 10^8 \, \mathrm{m\,s^{-1}}$?

4 What is the period of the wave in question 3?

5 Violet light has wavelengths between 400 nm and 450 nm. What frequency range does this account for?

6 Orange light covers the frequency range from 4.8×10^{14} Hz to 5.0×10^{14} Hz. What wavelength range is this?

7 What is the difference between a transverse and a longitudinal wave? Give an example of each.

8 What type of wave is a light wave?

9 Copy and complete Figure 11.9, labelling all angles and lines. What does this diagram demonstrate?

Figure 11.9

10 With the aid of a diagram, explain how the curved reflector of a satellite receiver operates.

Interference of coherent waves

When two waves of the same type interact, there will be some form of interference. To investigate this, we introduce a new term – **phase**. Phase allows us to compare the point in the cycle that a wave has reached but because it must be measured with reference to another wave, we talk of **phase difference**.

Two waves which are identical in terms of frequency, wavelength and velocity may have a different phase. This is shown in Figure 11.10 and means that the waves will be at a different stage in their cycle at a given point in space and time. If the relationship between the two waves, the phase difference, is constant, then these two waves can be said to be **coherent**.

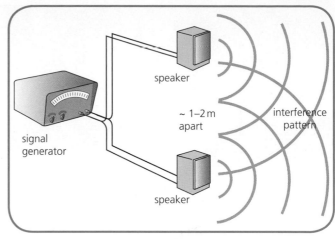

Figure 11.11 Demonstrating interference with sound waves. If you move around the room with the speakers set up as shown, there will be quiet and loud areas due to the interference of the sound waves from each speaker

| **For Interest** | **Coherent waves** |

Waves are said to be coherent when they have the same frequency, wavelength and velocity and also a constant phase difference. A constant phase difference means that the difference in position between successive waves remains the same all the time.

A simple demonstration of phase difference can be set up using two speakers. The speakers should be producing exactly the same sound, but should be positioned at different points in a room. The waves spread out throughout the room and interact with each other as shown in Figure 11.11.

This experiment could equally be performed with a simple sound such as that from a signal generator. Walk around the room and listen closely at different points.

You could use a sound level meter. You need to make sure you stand far enough away from the speakers for the waves to be interacting.

If two coherent waves interact they interfere. We describe **interference** as **constructive** and **destructive**. At certain points waves (which are coherent) interact in such a way and produce maxima and minima. Figures 11.12 and 11.13 show the effects of constructive and destructive interference where waves can 'increase' or 'decrease' depending on the phase difference.

Constructive interference is generally described in terms of two coherent waves coming together at their maxima (positive or negative), resulting in a wave with double the amplitude of the original wave. This occurs when there is no phase difference between two waves or the phase difference is a

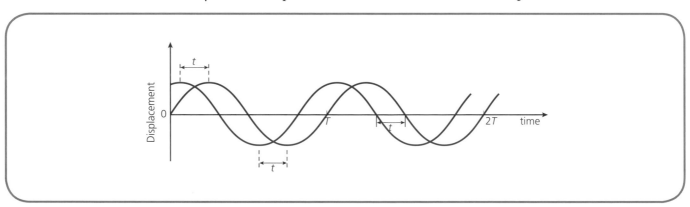

Figure 11.10 The phase difference is a measure of how far apart the cycle of two waves is. It can be measured in time, according to the period, or more commonly in distance, according to the difference in wavelength

whole number multiple of the wavelength. More correctly, constructive interference occurs when the amplitude of the combined waves is greater than that of the individual waves. It need not be crest and crest combining but these are the most obvious and noticeable occurrences.

Conversely when the phase difference is exactly one half wavelength, the maximum of one wave will meet the minimum of the other. As these waves are the same, the amplitude of the second wave is the same magnitude as that of the first but in the opposite direction, so they combine and effectively cancel each other out. It is important to note that though the two waves can cancel each other out, this effect only happens at a very specific point – both waves are still there and will continue past the interference point. More correctly, destructive interference occurs when the amplitude of the combined waves is less than that of the individual waves. It need not be crest and trough combining but these are the most obvious and noticeable occurrences.

The two waves shown in Figure 11.14 overlap at various points causing an interference pattern where there are multiple maxima and minima.

We can use the relationships at maxima and minima to calculate how much further one wave has travelled than the other. The difference in distance between the paths of the two waves is called the **path difference**. It is measured in metres. If a wave is constructively interfering, and creating a maximum, the path difference must be a whole number of wavelengths.

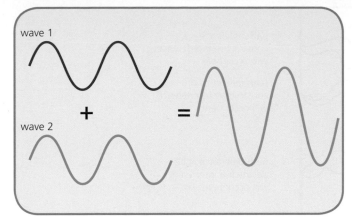

Figure 11.12 As the maxima of these two waves occur at the same time, they will interfere constructively. The resultant wave will have double the amplitude of each single wave

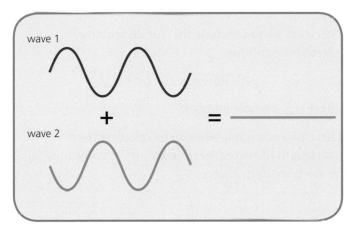

Figure 11.13 The maximum of one wave occurs at the same point as the minimum of the other. This results in destructive interference – the waves cancel each other out. This effect only happens at the exact point where the path difference between two waves is $\lambda/2$. The waves both still exist and will continue after this point

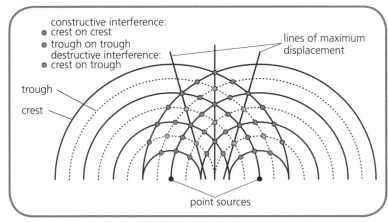

Figure 11.14 There will be multiple points of constructive and destructive interference depending on the path difference at each point

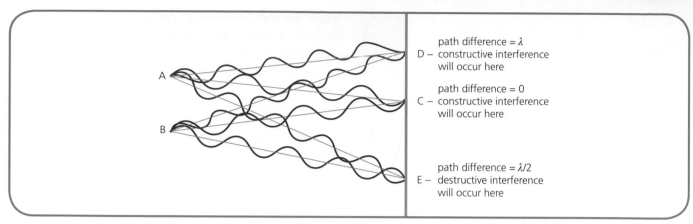

Figure 11.15 Diagram of waves combining and interfering

This leads us to the conclusion that for constructive interference maxima

$$\text{path difference} = n\lambda$$

where n must be a whole number and denotes the number of wavelengths between the two waves.

If the waves are interfering destructively and are creating a minimum, this must mean that the waves are half a wavelength different in phase. As a result, we

know that the path difference must be an odd multiple of half wavelengths: ½, 1½, 2½, 3½ and so on.

This leads us to conclude that for destructive interference minima

$$\text{path difference} = (n + \tfrac{1}{2})\lambda$$

where n is a whole number.

These two equations allow us to calculate the wavelength of two coherent waves if we know how far we are from each source.

Worked Example 11.2

Two coherent waves are produced. They are separated by a distance as shown in Figure 11.16.

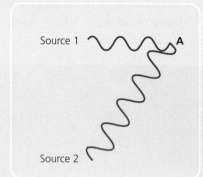

Figure 11.16

Consider the wave from point A which is horizontal to Source 1.

a) How many wavelengths is the path difference at point A on the diagram?

b) What is the path difference at point A if the wavelength is 50 cm?

c) What is the distance between the two sources?

a) From the diagram, you can see that point A is a maximum. It is three wavelengths from the first source and five wavelengths from the second source. So, the path difference is $5\lambda - 3\lambda = 2\lambda$.

b) Path difference = $n\lambda$

Path difference = $2 \times 0.5 = 1\,\text{m}$

→

c) You can use Pythagoras to calculate the distance between the sources. A right-angled triangle can be drawn as shown in Figure 11.17.

$$a^2 = b^2 + c^2$$
$$5 \times 0.5 = 2.5\,m$$
$$3 \times 0.5 = 1.5\,m$$
So $$2.5^2 = 1.5^2 + distance^2$$
$$6.25 - 2.25 = distance^2$$
$$distance = \sqrt{4} = 2\,m$$

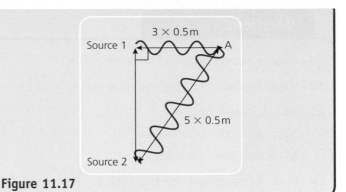

Figure 11.17

Interference in everyday life

You can see various examples of interference all around you. Often what we see is 'thin-film' interference which occurs when light is reflected from the top and the bottom of a thin film of a substance.

One example most people have seen is that of oil on water. A very thin layer of oil will reflect light from its surface, but also from the interface between the oil and water below it. These waves interfere with each other just like the waves we have looked at so far. In some places, this will create very bright areas of certain colours (where constructive interference has taken place for that wavelength of light) and in other places, certain colours will destructively interfere so we will not be able to see them at all.

A similar effect is seen when a bubble is in the air. In this case the light is reflecting from all of the boundaries between the air and the soapy water.

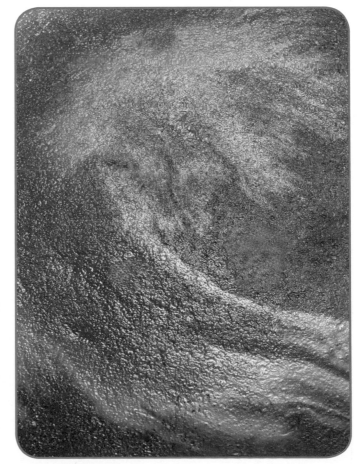

Figure 11.18 Interference produces the patterns we see in bubbles and oil slicks

Questions

11 What is the definition of coherence?

12 What does a constant phase relationship mean?

13 What is meant by
   a) constructive interference
   b) destructive interference?

14 In each case below, the two separate waveforms meet at a point. Copy the diagrams and draw the resulting waveform seen at that point.
   a)

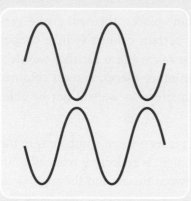

Figure 11.19

   b)

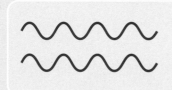

Figure 11.20

15 a) Which of these waves are coherent?
       i)

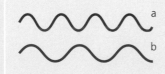

Figure 11.21

       ii)

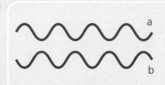

Figure 11.22

   iii)

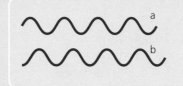

Figure 11.23

   iv)

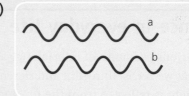

Figure 11.24

   b) What is the phase difference of the coherent waves in part a)?

   c) Which of the coherent waves will cause complete destructive interference?

16 Two light sources have interfered destructively. If a photo detector is placed at that point, what will the reading be?

17 What is the difference between 'phase difference' and 'path difference'?

18 Describe an experiment which can be used to demonstrate interference. Include details of the source, any detectors you would use and a diagram of the results.

19 Two coherent sources of light produce a minimum 3 m from one source and 1.5 m in front of the other. What is the path difference?

20 The first maximum of two coherent sources which are interfering occurs 3 m from one source and 4 m from the other. What is the wavelength of the sources?

21 Study Figure 11.25 and then answer the following questions for points a–f.
i) Is interference occurring in this diagram?
ii) If interference is occurring, is it constructive, destructive or varying?
iii) Is it a maximum, minimum or neither?

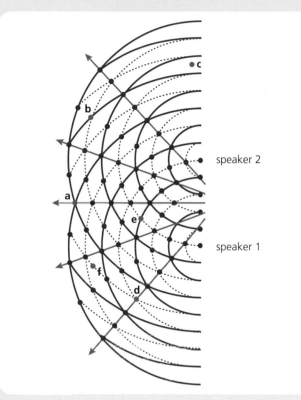

Figure 11.25

22 Calculate the wavelength of the following coherent waves.

a)

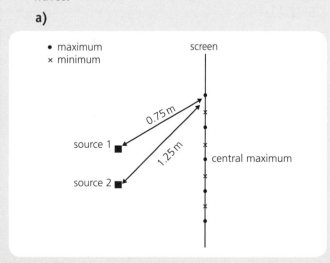

Figure 11.26

b)

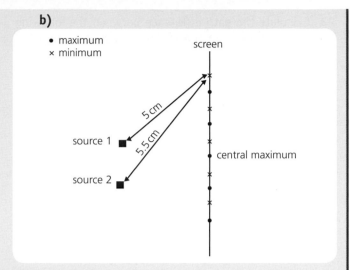

Figure 11.27

c)

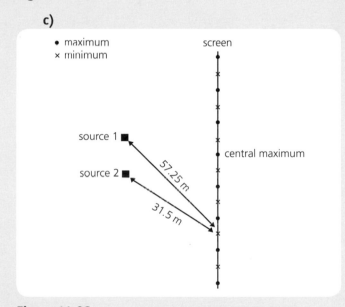

Figure 11.28

d)

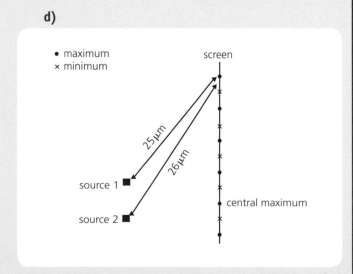

Figure 11.29

23 Calculate the path difference between the following coherent waves at the points indicated in red.

a)

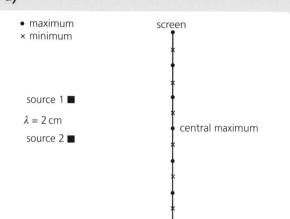

Figure 11.30

b)

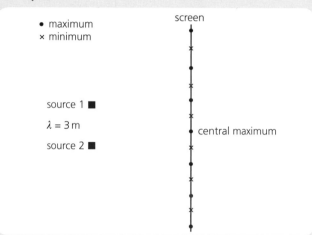

Figure 11.31

c)

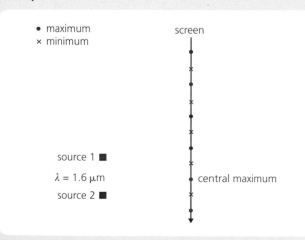

Figure 11.32

d)

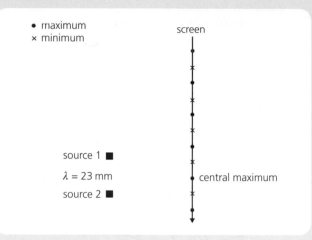

Figure 11.33

24 Calculate the missing path length in each case.

a)

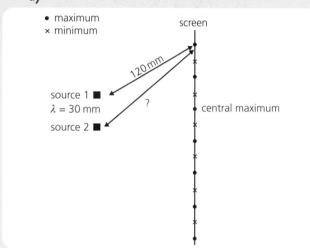

Figure 11.34

b)

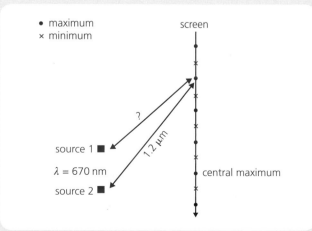

Figure 11.35

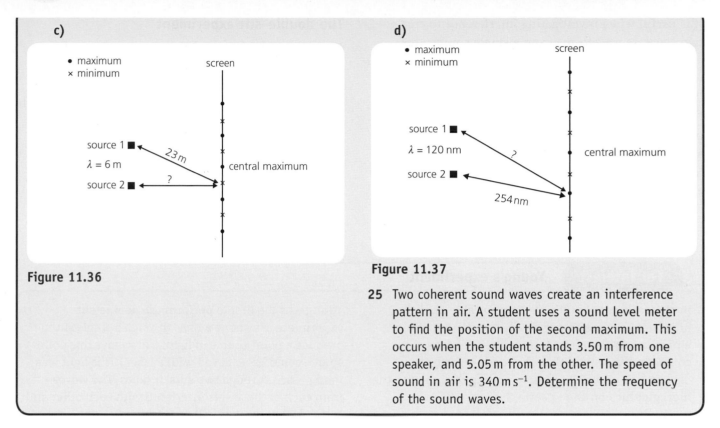

c)

• maximum
× minimum

screen

source 1 ■

λ = 6 m

source 2 ■

23 m

?

central maximum

Figure 11.36

d)

• maximum
× minimum

screen

source 1 ■

λ = 120 nm

source 2 ■

?

254 nm

central maximum

Figure 11.37

25 Two coherent sound waves create an interference pattern in air. A student uses a sound level meter to find the position of the second maximum. This occurs when the student stands 3.50 m from one speaker, and 5.05 m from the other. The speed of sound in air is 340 m s⁻¹. Determine the frequency of the sound waves.

Diffraction

When a wave passes an obstruction, for example travels around a hill or through a gap, the wave spreads behind the regions which were obstructed. This is why radio signals can be detected behind buildings and hills. This 'spreading' is referred to as **diffraction**. You will already have seen this effect in diagrams from the previous section where the waves spread out from the sources.

It is possible to investigate these effects using a ripple tank. If an obstacle with a gap is placed in front of a water wave, the wave will continue through the gap. You might expect the wave to continue in a straight line from the gap, but this does not happen. Instead, it spreads out into the area shielded by the obstacle. The spread depends on various aspects of the situation including the velocity, frequency and wavelength of the wave and also the width of the gap. We can see this effect when looking at waves lapping on the shore and diffracting around stones.

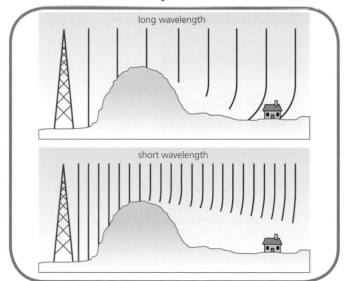

Figure 11.38 Waves will spread out around obstacles – this is called diffraction

Figure 11.39 The waves bend round the obstacle. The amount of diffraction depends on many different elements

It is useful when investigating interference to create two separate waveforms from a single source; this ensures coherence. One way of doing this is to have the wave pass through a mask or grating with two gaps in it. This will produce two separate waves with the same frequency, wavelength and velocity. On the other side of the mask these waves spread out due to diffraction as if they were two separate waves. This observation led to a famous experiment, called the double-slit experiment.

The double-slit experiment

Young's double-slit experiment has been extended since its development in the early nineteenth century and is now used in various investigations. Modern scientists use a laser as the coherent light source, which is directed to two slits in a mask. These are extremely close together and narrow. Figure 11.41 shows the interference pattern created when a red laser is used. Clearly seen maxima are produced where the intensity is greater and destructive interference causes gaps in the pattern where there is no light at all.

For Interest	Young's experiment

In the early nineteenth century, notable English physicist Thomas Young developed the experiment which bears his name. Interestingly he was considered a polymath as he made significant discoveries in many areas, including deciphering the hieroglyphics on the Rosetta Stone.

Young was the first to perform the double-slit experiment. He shone a light through a single slit to produce a point source of light. (Nowadays this point source could be achieved with a laser.) This light was then passed through two slits or holes. The waves from each of these slits interfered with each other and created what we now call an **interference pattern**.

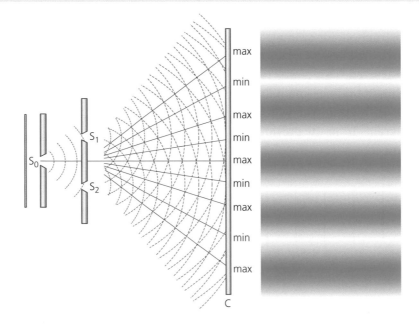

Figure 11.40 Young's original experiment used a basic light source and so a single slit was required to make a small 'point' light source. This is not required when using a laser, as the laser is already an approximation of a point source

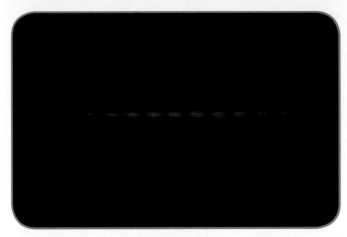

Figure 11.41 The interference pattern caused when a red laser is aimed through a double slit

It is possible to calculate the distance between the slits based on the position of maxima and minima. We assume that the screen on which the interference pattern is projected is some distance from the slits so that the beams from each slit are almost parallel.

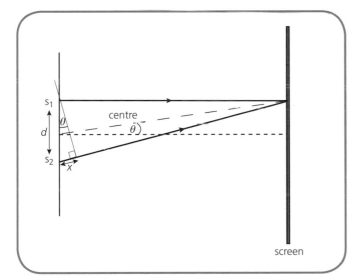

Figure 11.42 Here we see two parallel beams leaving a double slit. If the screen is far enough away, the two beams will be almost parallel at the maxima. This allows us to make an approximation that the two angles shown are the same size – as they will be almost equal. Using this approximation we can calculate x which is the path difference

Figure 11.42 shows the angles involved. As the rays are almost parallel, we can conclude that the two angles shown as θ are approximately the same. Using trigonometry we can state that the sine of the angle θ equals the path difference divided by the slit separation.

In Figure 11.42, the path difference is denoted as x and the slit separation as d. We know that for maxima the path difference must be a whole number of wavelengths, so for any angle which leads to a maximum, we replace x with $m\lambda$ where m is a whole number.

We then have the equation $\sin\theta = m\lambda/d$ which we tend to write as

$$d\sin\theta = m\lambda$$

where θ is the angle from the centre of the grating to a maximum, d is the distance between slits, m is a whole number and λ is the wavelength of the coherent sources.

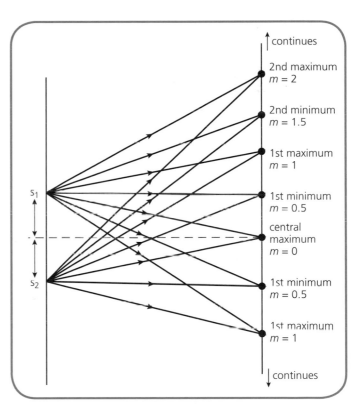

Figure 11.43 An interference pattern is also created causing maxima and minima to occur

This equation holds for multiple slits too. A larger number of slits makes the maxima more defined and easier to determine. In this case d refers the distance between two consecutive slits.

Worked Example 11.3

A grating is placed in front of a laser emitting light of wavelength of 635 nm. The third maximum occurs at an angle of 27° to the centre of the grating. Calculate the distance between the slits.

We know that angle $\theta = 27°$, the wavelength is 635 nm and m is 3.

We use the relationship $d \sin\theta = m\lambda$

$d \sin 27° = 3 \times 635 \times 10^{-9}$

$d = 1.905 \times 10^{-6}/\sin 27° = 1.905 \times 10^{-6}/0.45399$

$d = 4.1961 \times 10^{-6}$ m

Therefore the distance between the slits is 4.2 μm.

Research Task

How has the double-slit experiment impacted on particle physics?

What happens when a series of individual electrons are passed through a double slit instead of light? What does this mean for electrons and light?

Grating

If we increase the number of slits used in these experiments, the spectra produced become clearer than with just two slits. Slides with 3, 4, 5 (…) slits make the maxima more defined.

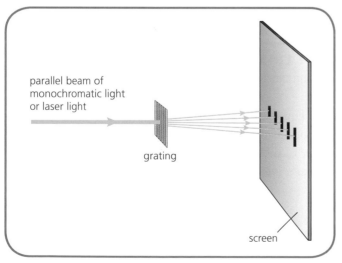

parallel beam of monochromatic light or laser light

grating

screen

Figure 11.44

Gratings are specially-fabricated pieces of glass with a great many slits, which make observations possible in classroom laboratories.

A white light source may also be used with a grating. When a grating is placed in front of a thin beam of white light, spectra will be produced on the screen as opposed to a 'normal' interference pattern. This illustrates that the interference observed is a function of wavelength. White light is made up of a number of different wavelengths (colours). Recall that the condition for a maximum is:

$$d \sin\theta = m\lambda$$

This results in individual interference patterns (maxima and minima) for each colour of light. The central maximum will remain close to white light but either side of the central maximum will be spectra which spread into each colour. The position of each colour in this spectrum is representative of the spread of wavelengths. The spectra will be symmetrical around the central maximum so the colours on the left will be in the opposite order to those on the right.

Interferometers

Interference patterns have found useful applications in microscopy. Using different wavelengths of electromagnetic waves it is possible to use the path difference to calculate the depth of an object. The waves are moved over the surface of an object and a detector monitors the maxima and minima collected from their reflections. The spacing of the interference pattern depends on tiny changes in the surface of the object which is being investigated. The changes can then be interpreted by a computer to calculate the change in depth from one point to another compared to an original reference. These machines are particularly useful for creating a detailed image of small features which are being manufactured on a product such as microfluidic channels and to ensure that surfaces are perfectly flat such as mirrors for use in space telescopes.

Research Task

Can you find any other uses of interferometry? Find out more details of how it works. How else can interference be used to identify features?

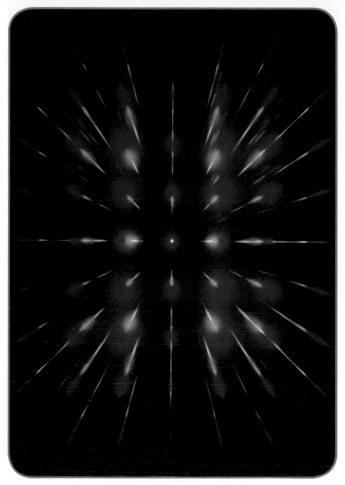

Figure 11.45 A diffraction pattern

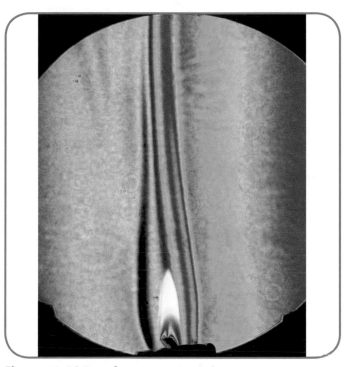

Figure 11.46 Interference pattern shown on an interferometer image of a flame

Questions

26 What is meant by monochromatic light?

27 A student wants to create a full visible light spectrum. Describe how they would do this using a grating.

28 Explain, with the aid of a diagram, the approximations made in reaching the relationship $d \sin\theta = m\lambda$.

29 In the following examples, calculate the wavelength of the source, assuming the sources are coherent.

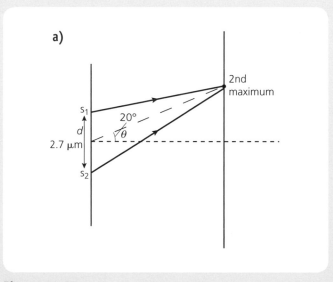

Figure 11.47

30 In the following examples, calculate the distance between the slits, assuming the sources are coherent.

b)

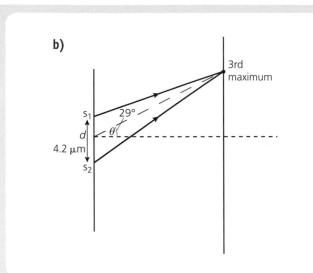

Figure 11.48

a)

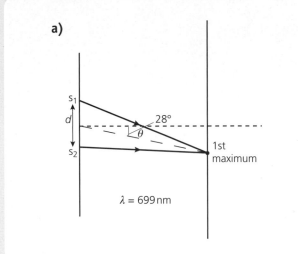

Figure 11.51

c)

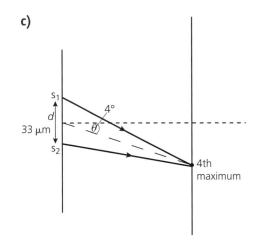

Figure 11.49

b)

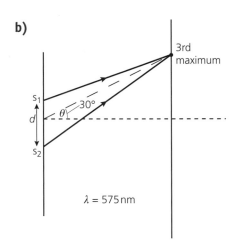

Figure 11.52

d)

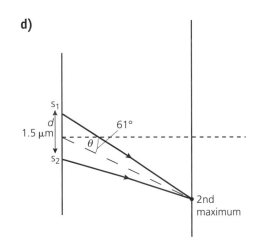

Figure 11.50

c)

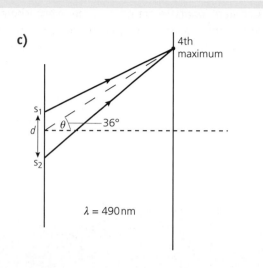

Figure 11.53

d)

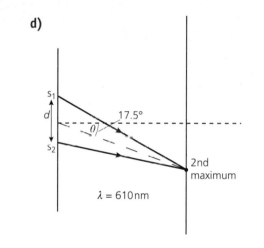

Figure 11.54

31 In the following examples, calculate the angle between the centre of the grating and the highlighted maxima, assuming the sources are coherent.

a)

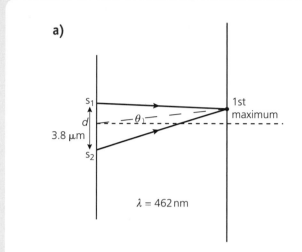

Figure 11.55

b)

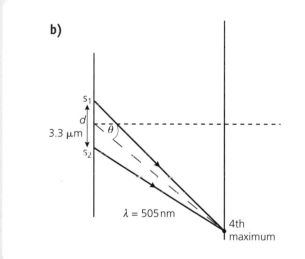

Figure 11.56

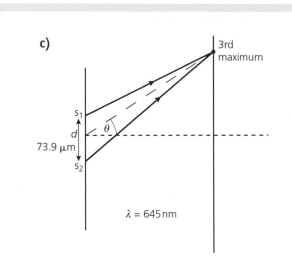

Figure 11.57

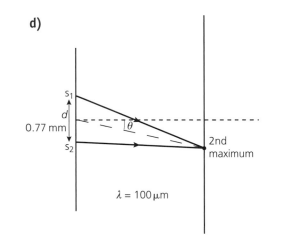

Figure 11.58

32 A student wants to create an experiment where he has three maxima on either side of the central maximum but the maxima must fit within 45° either side of the grating. The only light source available is a blue 470 nm laser. What maximum grating spacing is required to achieve the desired result?

33 Gratings can be used to produce interference patterns horizontally and vertically at the same time, as shown in Figure 11.59.

Figure 11.59

Using your knowledge of how a single interference pattern is created, explain how you think the pattern in the picture was generated.

Consolidation Questions

1 Two fire-alarm sirens are sounding during a test. They are identical and their outputs are in phase at a frequency of 1.6 kHz. They are positioned 5.2 m apart from each other at the corners of a building.

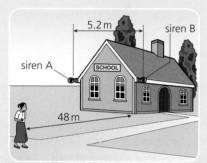

Figure 11.60

A person is walking along a road 48 m away. The person detects a series of maxima and minima, similar to the situation where maxima and minima of light intensity are observed in a double-slit experiment.

a) With reference to path difference explain the detection of
   i) maxima of sound
   ii) minima of sound.

b) Each siren also has a bright lamp which gives a visual warning. State why no interference pattern is observed by the person as they walk past the alarms. You must justify your answer.

2 The apparatus shown in Figure 11.61 below is set up to determine the wavelength of light from a laser.

The wavelength of the light is calculated using the equations

$$\lambda = d \sin\theta \text{ and } \sin\theta = x/L$$

The angle θ, length L and separation distance are indicated in the diagram.

a) The distance L is measured by 8 students. The measurements are: 2.404 m, 2.401 m, 2.416 m, 2.401 m, 2.402 m, 2.392 m, 2.419 m, 2.388 m

   Determine the mean value of L and the approximate random error in the mean.

b) The measurement of the separation x is (92 ± 2) mm. Determine whether L or x has the greatest percentage error.

c) Determine the wavelength, in nanometres, of the laser light. Your answer should include the uncertainty.

d) Using the same equipment and with no increase in the number of measurements, suggest an improvement in the experimental setup.

3 Monochromatic light is incident normally upon a grating which has 350 lines per mm. The angle between the two second order maxima is 39° as shown in Figure 11.62.

a) Calculate the wavelength of the monochromatic light.

b) Determine the colour of this light.

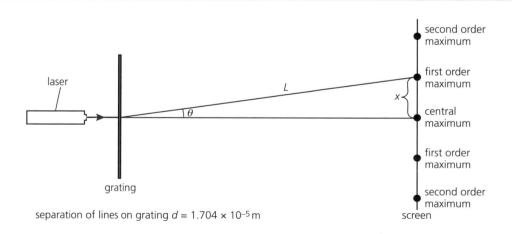

separation of lines on grating $d = 1.704 \times 10^{-5}$ m

Figure 11.61

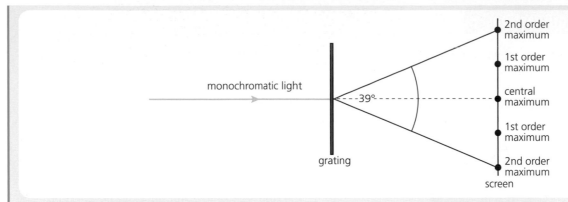

Figure 11.62

4 An experiment to determine the wavelength of a microwave source is set up as shown in Figure 11.63.

   a) The detector is moved from A to B. The reading on the microwave detector increases and decreases several times. State, in terms of waves, how the pattern of maxima and minima is produced. You must explain your answer.

   b) The measurements from the metal plate to the detector are measured at the third order maximum as shown. Determine the wavelength of the waves.

5 a) An experiment requires laser light with wavelength 642 nm to be incident on a grating. A series of bright spots are seen on a screen placed a significant distance away from the grating. The distance between these spots is shown in Figure 11.64.

      Calculate the number of lines per metre on the grating.

   b) The laser is replaced with a new laser. The experiment is repeated and the bright spots appear closer together. State how the wavelength of the new laser compares to that of the original laser. You must justify your answer.

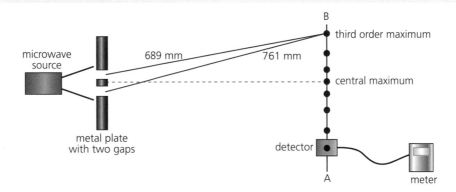

Figure 11.63

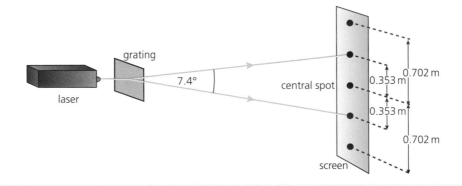

Figure 11.64

12 Refraction of light

When a wave enters a new medium, the velocity of that wave may change. Unless the entry to the new medium is perpendicular to the surface of the new medium, this will result in a change in direction. We describe the phenomenon of the change in direction of a wave as a result of the change in velocity as **refraction**.

You can perform a very simple experiment to see refraction. Half fill a glass with water and then place a straw in it. If you look at the straw from above, it will appear to bend. From the side it may look like it has split in half!

Figure 12.1 Refraction occurring in water

In this example there are actually three materials – air, glass and water – each of which has a different **refractive index**. The refractive index is a measure of the relationship between the velocities in two materials. To gain a clearer understanding of how it comes about we must investigate the angles involved in refraction and for this purpose we will look mainly at light waves. While refraction occurs for all waves, we will focus on light waves as they are visible. To begin, we will look at a single change of medium – light entering a glass block (see Figure 12.2).

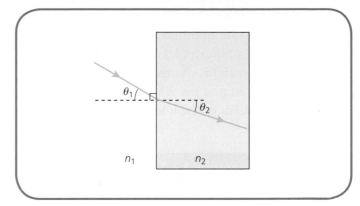

Figure 12.2 Refraction in a glass block. When the light enters the new medium it changes velocity and as a result it changes direction

The light enters the glass block at an angle to the normal. We define this as the **angle of incidence**. The light then changes direction at the point of incidence and continues at a different angle, the **angle of refraction**.

We can state the relationship between the angles in Figure 12.2 as

$$n_1 \sin\theta_1 = n_2 \sin\theta_2$$

This can be rearranged to give

$$n_2/n_1 = \sin\theta_1/\sin\theta_2$$

The letter n denotes the refractive index of the medium. The refractive index of any given medium is defined as the ratio of the speed of light in a vacuum (c) divided by the speed of light in that medium

$$n = c/v$$

At this stage we tend to assume that the refractive index of air is 1 (although in reality, it is slightly above 1). This assumption is unlikely to make a noticeable difference to the angle of refraction measured in schools. This leads us to the definition of refractive index. We define the refractive index in a vacuum as 1 and use it to state all other refractive indices.

It is important to understand how the refractive index affects a light ray as it enters a glass block. We use the definition of refractive index along with the ratio outlined earlier.

If the refractive index of air is 1, then $n_1 = 1$ and the equation becomes

$$n_2 = \sin\theta_1/\sin\theta_2$$

or, as it is usually written,

$$n = \sin\theta_1/\sin\theta_2$$

This equation is only true when one of the angles is in a material with a refractive index of 1.

Shining a single beam of monochromatic light into the block shown in Figure 12.2 allows us to investigate Snell's law. If θ_1 (in air) is changed and θ_2 (in the block) is measured, we can plot $\sin\theta_1$ (y-axis) against $\sin\theta_2$ (x-axis). The gradient of the line will be the refractive index of the block.

Worked Example 12.1

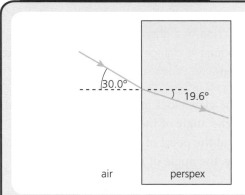

Figure 12.3

Calculate the refractive index of the plastic block shown in Figure 12.3.

The light in air has an angle of 30.0° so $\theta_1 = 30.0°$

The light in the plastic block has an angle of 19.6° so $\theta_2 = 19.6°$

Since the angle of incidence is in air, we can use the relationship $n = \sin\theta_1/\sin\theta_2$

$n = \sin\theta_1/\sin\theta_2 = (\sin 30.0°)/(\sin 19.6°) = 1.49$

It is important to note that when using this particular equation, the angle θ_1 must always be the angle in air. It does not matter whether it is the angle of incidence or refraction. For example, if we extend our earlier diagram and show the ray of light exiting the glass block, the angle will become greater (Figure 12.4).

We can see that the ray of light leaving the glass block is parallel to the ray which entered the block. This is because the inverse (opposite) change in direction occurs as the ray leaves the block.

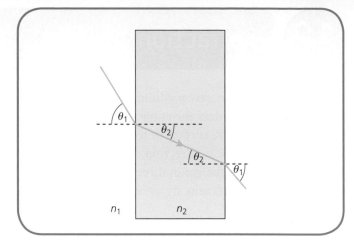

Figure 12.4 Refraction in a glass block. The light changes direction at the point at which it changes medium. The same occurs as it moves from glass back to air, but the change in direction is reversed, so the light rays entering and leaving the glass block are parallel

The frequency of a light wave is set by the source of that wave. When the wave enters a different medium, the frequency must stay the same – in other words, the number of peaks and troughs passing through the medium must continue – otherwise they would start to 'stack up' and push into one another. Consider the wave equation $v = f\lambda$. We have already discovered that when a light wave enters a new material, its velocity changes. If the frequency stays the same when the wave enters the new material, then according to the wave equation, the wavelength must change.

If we make f the subject of the wave equation, we have

$$f = v/\lambda$$

The frequency before entering the new medium equals the frequency after so

$$v_1/\lambda_1 = v_2/\lambda_2$$

and we can rearrange this to give the relationship

$$v_1/v_2 = \lambda_1/\lambda_2$$

This relationship can be extended further when we consider the earlier relationship $n_2/n_1 = \sin\theta_1/\sin\theta_2$

The change in angle is due to the change in refractive index, which is in turn due to the change in velocity, so we can state

$$\sin\theta_1/\sin\theta_2 = \lambda_1/\lambda_2 = v_1/v_2$$

Any part of this relationship can be used to calculate the change in wavelength, velocity or direction as a result of refraction at the interface between two materials.

Worked Example 12.2

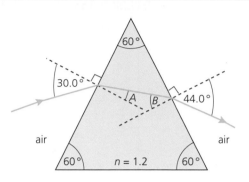

Figure 12.5

Fill in the missing angles in Figure 12.5.

The first thing to note is the refractive index. It is stated as $n = 1.2$.

The next stage is to work out angle A. The angle of

incidence from air is 30° and we can use the equation from earlier

$n = \sin\theta_1/\sin\theta_2$
$1.2 = (\sin 30.0°)/(\sin A)$
$\sin A = (\sin 30.0°)/1.2 = 0.5/1.2 = 0.416...$
$A = \sin^{-1}(0.416...) = 24.6°$

For angle B we must remember that the angle in air must always be θ_1 in the relationship $n = \sin\theta_1/\sin\theta_2$.

From this we say that

$\theta_1 = 44.0°$ and $\theta_2 = B$

n remains the same so

$n = \sin\theta_1/\sin\theta_2$
$1.2 = (\sin 44.0°)/(\sin B)$
$\sin B = (\sin 44.0°)/1.2 = 0.694.../1.2 = 0.578...$
$B = \sin^{-1}(0.578...) = 35.4°$

Dependence on wavelength

The refractive index of a material is not constant for all frequencies of light. We can say that it is dependent on the frequency or wavelength of the wave in a vacuum. As a general rule, the longer the wavelength, the lower the refractive index. This change in refractive index is the reason that prisms can split white light into its constituent parts. Red has the longest wavelength, so the refractive index of a material is less for red light than for violet light, which has the shortest wavelength.

Figure 12.6 The refractive index of a material is dependent on the wavelength of light. This is why a prism can disperse white light into different colours – as the refractive index varies depending on the colour

This is similar to the effect which causes rainbows. Water droplets in the air act in the same way as the prism by refracting the light and thus separating it into its constituent colours. The name given to this separation of colours is **dispersion**.

In Chapter 11 we showed how it is possible to create spectra using gratings. There are a few differences between the two methods of making spectra and these are detailed below.

- Gratings rely on diffraction and interference while prisms rely on refraction.

- Prisms will tend to produce a single spectrum from a single ray of light (more may be visible depending on various reflections). When spectra are produced using gratings, however, there will be a number of spectra caused by the interference pattern of maxima and minima.

- The order of the colours produced by a prism are in the reverse order to that produced by a grating. The deviation from the central line is least for red for the prism and greatest for the grating.

Questions

1 Explain the phenomenon of refraction. A diagram may assist your explanation.

2 How is refraction related to dispersion?

3 In the examples below, calculate the refractive index of the block.

a)

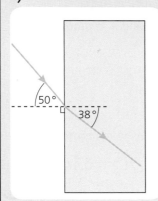

Figure 12.7

b)

Figure 12.8

c)

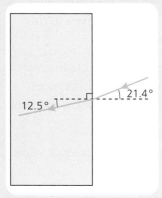

Figure 12.9

d)

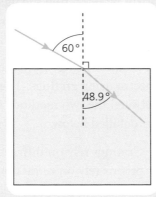

Figure 12.10

4 Copy and complete Table 12.1. In all cases, light is entering a rectangular glass block from air.

5 In the examples below fill in the missing angle *x*.

a)

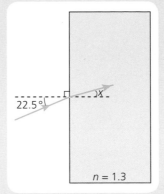

$n = 1.3$

Figure 12.11

b)

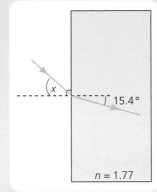

15.4°

$n = 1.77$

Figure 12.12

c)

29.7°

$n = 1.42$

Figure 12.13

d)

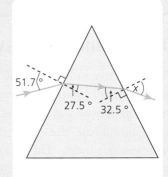

51.7°

27.5°   32.5°

Figure 12.14

6 A beam of light of frequency 461.5 THz is directed at a glass block which has a refractive index of 1.54. Calculate:

a) the wavelength in air

b) the wavelength in the block

c) the velocity in the block

d) the angle of refraction if the angle of incidence is 22°.

Refractive index	Angle of incidence/°	Angle of refraction/°	λ_1/nm	λ_2/nm	v_1/m s^{-1}	v_2/m s^{-1}
A	B	6.2	675	C	3.0×10^8	1.3×10^8
1.21	73.2	D	697	E	F	2.5×10^8
1.64	54.3	G	H	331	I	J
K	L	15.3	457	243	M	N

Table 12.1

7 A beam of light meets a large triangular block of plastic with a refractive index of 1.2. The triangle is equilateral and the light meets the centre point of one side at an angle of 15.0°. Draw an accurate diagram showing the path taken by the light until it exits the block. (Show all angles.)

8 A 4.0 cm cubic glass block is used to change the path of light in an experiment. The refractive index of the glass block is 1.5. When positioned correctly, the light exits the glass block 3.0 cm higher than where it entered.

a) Determine the angle required to displace the beam by 3.0 cm.

b) Draw a scale diagram showing all angles required for this experiment to work.

9 Refractive index is not constant. Explain this statement, including reasons.

10 Describe two methods of producing a full visible light spectrum and explain how the methods differ.

Total internal reflection and the critical angle

When we wish to send a large amount of data over a long distance we need an inexpensive and efficient system. Older communication systems used copper cables but these need a lot of energy to maintain signal strength, because even high-purity copper has a small resistance. They are also usually only able to take a single signal per wire, they are heavy and they have become very expensive because of the basic price of copper.

Many modern systems use fibre optic cable. Thin strands of glass can be bunched together to make a single cable. Each cable can be used for multiple signals and each signal requires significantly less power.

In order to understand how a signal can be sent so efficiently using fibre optic cable, we need to look at the optical properties of the cable. When a light ray reaches the boundary of a material it will often refract and reflect, as shown in Figure 12.16. If we increase the angle of incidence, the angle of refraction increases. As we continue to increase it, the angle of refraction will continue to get larger until it reaches exactly 90° as shown in Figure 12.17 on the next page.

Figure 12.15 The total internal reflection occurring in a fibre optic cable allows signals to travel long distances at a fraction of the price of copper cables

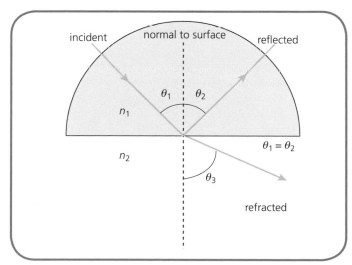

Figure 12.16 The light entering this block has a refracted and a reflected ray. The angle of incidence is less than the critical angle

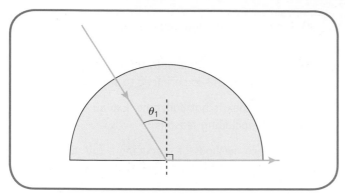

Figure 12.17 The angle of incidence is now at exactly the critical angle. At this point the angle of refraction is exactly 90°

You can try this with a simple block of glass and a single light ray. If you start with a narrow angle of incidence and steadily increase it, you will reach the point where the angle of refraction is exactly 90 degrees. This is often difficult to see but you should notice that there is some reflection throughout.

If you continue to increase the angle of incidence, you will notice that there is no longer any refraction – but the reflection remains. Any angle of incidence greater than the so-called **critical angle** will result in reflection without refraction.

When the angle of incidence is greater than the critical angle all light is reflected internally and we refer to this as **total internal reflection**.

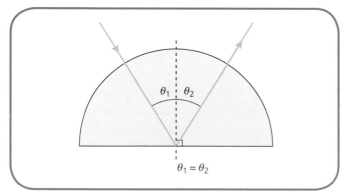

Figure 12.18 The angle of incidence in this case is greater than the critical angle, so there is no longer an angle of refraction. All of the light is internally reflected and we call this total internal reflection

It is this total internal reflection that allows us to use fibre optic cables for communication. Provided the angle of incidence of light remains greater than the critical angle, all of the light will be reflected and will

continue to travel through the fibre optic cable until it reaches the end. At the end point, it is detected and converted back to an electrical signal to be used in a circuit. Figure 12.19 shows how this occurs.

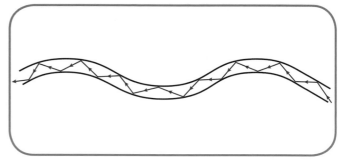

Figure 12.19 Total internal reflection happens repeatedly down a fibre optic cable allowing the signal to travel to its destination. It is important that the cable is not bent too much though

Research Task

In what other ways are fibre optic cables used? Are there any applications in medicine? How can a single strand of glass transmit more than one signal? What methods can be used?

It is possible to investigate the critical angle of various materials experimentally, but if you know the refractive index of a material, you can work it out by a simple calculation.

The relationship between critical angle and refractive index is

$$\sin\theta_c = 1/n$$

where θ_c is the critical angle and n is the refractive index of the material.

Worked Example 12.3

A block of glass has a refractive index of approximately 1.5. What is its critical angle?

$\sin\theta_c = 1/1.5$

$\theta_c = \sin^{-1}(0.67)$

$\theta_c = 41.8° = 42°$

Research Task

The concept of total internal reflection is part of why diamonds sparkle.

Find out about their refractive index and critical angle and try to explain why they are so carefully cut.

Figure 12.20 What is it about a diamond's refractive index that makes it so sparkly?

Consolidation Questions

1 What is meant by the term 'critical angle'?

2 Explain the term 'total internal reflection'. Include at least one diagram as part of your answer.

3 Calculate the critical angle if the refractive index is
   a) 1.26
   b) 2.41
   c) 1.87.

4 Calculate the refractive index of a material if the critical angle is
   a) 37.3°
   b) 72.2°
   c) 45.8°.

5 In each of the figures below, calculate the critical angle and state whether or not total internal reflection will occur.

a)

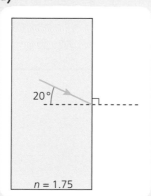

Figure 12.21

b)

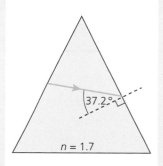

Figure 12.22

c)

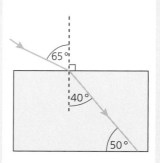

Figure 12.23

d)

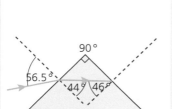

Figure 12.24

6 Fibre optic cables are often used for communications. Draw a diagram which represents the path of light down a fibre optic cable.

7 Why do fibre optic cables carry information more efficiently than copper cables?

8 Redraw Figures 12.25 to 12.28, completing the path of light so that it enters and exits the glass block.

a)

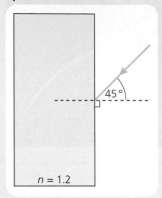

$n = 1.2$

Figure 12.25

b)

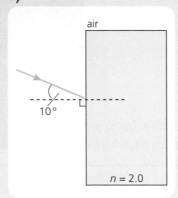

air

$n = 2.0$

Figure 12.26

c)

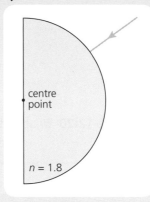

centre point

$n = 1.8$

Figure 12.27

d)

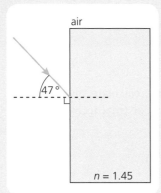

air

$n = 1.45$

Figure 12.28

The photoelectric effect and wave particle duality

The **photoelectric effect** was first described by Heinrich Hertz in the late 1800s as he was experimenting with the production of radio waves using a spark-gap apparatus.

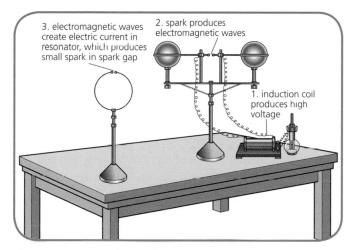

3. electromagnetic waves create electric current in resonator, which produces small spark in spark gap

2. spark produces electromagnetic waves

1. induction coil produces high voltage

Figure 13.1 Hertz's spark-gap apparatus

The two loops of brass were complete circuits except for a small air gap between the brass spheres at either end of the loops. Based on the work of James Clerk-Maxwell, Hertz reasoned that a spark jumping the gap in one loop (the transmitter) would cause electromagnetic radiation to be produced and that this radiation could be detected at some distance by the second loop (the receiver). If a spark could be observed at the gap in the receiver, then the radiation must be transmitting energy between the two loops. The experiments worked extremely well. Hertz detected radiation up to 15 metres away and in subsequent work showed that the radiation was reflected, refracted and polarised. The biggest problem for Hertz was seeing the spark at the receiver, so he then tried the experiments in the dark. He recorded that placing the receiver in a dark case had the effect of decreasing the maximum spark length and that removing the sides of the case had no effect until he removed the side between the transmitter and receiver.

This piqued his interest and a subsequent series of experiments showed that the effect was dependent on the wavelength of light which was falling on the receiver. Comparing the effects of wood, glass and quartz shields on the spark length showed that quartz shielding meant a longer spark. He determined that the quartz shield was allowing ultraviolet light to fall on the spark gap and that this had the effect of increasing the spark length. Hertz himself was unable to account for the observed phenomena and it was 1899 before J.J. Thomson identified that the ultraviolet light was causing electrons to be emitted from the surface of the metal. He enclosed a polished metal disc in a vacuum tube and exposed it to ultraviolet radiation. Similar to his work on cathode rays, he was making the metal disc the cathode. However, in this case, the electrons were being ejected from the cathode by the ultraviolet radiation as opposed to the heating and electric field he had used previously. At this point, a model of what was happening was beginning to become apparent.

The atoms in the metal contained electrons whose kinetic energy could be increased by heating, in the case of cathode ray tubes, or by the oscillating electric field of the ultraviolet radiation. With sufficient energy, they could break loose from the attraction of the nucleus of the atom and then be accelerated away from the metal surface by an externally applied electric field. In the case of incident light, several effects were predicted:

- Increasing the intensity of the incident light should cause an increase in the kinetic energy of the electrons and so they should be emitted with greater average velocity.

- Increasing the frequency of the incident light should also increase the average velocity of the emitted electrons.

- For very low levels of incident light, it should take some time for an electron to gain sufficient kinetic energy to be emitted.

In 1902, Philipp Lenard (Hertz's research assistant) went on to perform the first detailed experimental investigation to determine the complete nature of what was then being called the photoelectric effect. He used the apparatus depicted in Figure 13.2.

Lenard's light source was an arc light which had a huge range of intensity and could deliver specific colours, in other words frequencies, of light. By making the illuminated metal the cathode (negative electrode) of an electric circuit and introducing a second electrode to collect the ejected electrons (the collector), he could use a sensitive ammeter in the circuit to measure the size of the current flowing when the light illuminated the metal under investigation. Further, using an external voltage to hold the collector at a variable negative electrical potential compared to the cathode, he could alter the size of the potential difference between the two metals and so determine the maximum kinetic energy of the ejected electrons. Finally, he used different types of metal in different experiments to examine the impact of changing the type of metal emitting the electrons. His results contained a few surprises.

- There was a measurable minimum potential difference which stopped all electrons reaching the collector and this did not depend on the irradiance of the illuminating light.

- Doubling the intensity similarly doubled the number of electrons emitted (in other words, the current measured) but had no effect on the kinetic energy of the emitted electrons.

- The maximum kinetic energy of the electrons emitted depended on the frequency of light being used to illuminate the metal – higher frequency meant higher kinetic energy.

- There was a minimum value of light frequency below which no electrons were emitted.

- There was no delay in the emission of electrons. Even for very low light levels, electrons were emitted immediately if the frequency was sufficiently high.

While Lenard did not explain these observations and their clash with the predicted results, the experimental work was of sufficiently high quality to win him the Nobel Prize for Physics in 1905. It was Einstein who went on to propose a theory to account for the experimental work of Lenard. In fact, it was this theory which led to Einstein's own Nobel Prize for Physics in 1921.

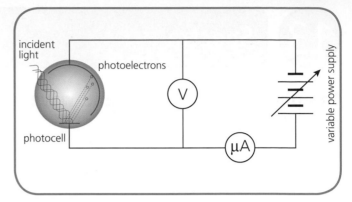

Figure 13.2 Photoelectric detector

His paper 'On a Heuristic Point of View Concerning the Production and Transformation of Light' was actually published in 1905, but it was so controversial that experimental proof by Robert Millikan was required before the award was made. (Millikan, an experimental physicist from America, spent a great deal of time trying to disprove Einstein's work but eventually admitted, 'I spent ten years of my life testing that 1905 equation of Einstein's, and, contrary to all my expectations, I was compelled in 1915 to assert its unambiguous experimental verification in spite of its unreasonableness since it seemed to violate everything that we knew about the interference of light.')

Einstein proposed the following:

- Light is quantised. In other words, it travels in discrete 'chunks' called **photons**. These photons were essentially particles of light.

- The energy of the photon is given by the equation

$$E = hf$$

where

E = energy (joules, J)

h = Planck's constant (J s)

f = frequency of the incident light (hertz, Hz)

- Like a billiard ball, the incoming light particle strikes the electron in its orbit around the atomic nucleus. If it has sufficient energy the photon will eject the electron. If not, the electron stays in place. Increasing the illuminating light irradiance will increase the rate of arrival of photons but not their individual energies.

- By conservation of energy, the equation for the photoelectric effect is:

$$hf = hf_0 + E_k$$

where

E_k = kinetic energy of ejected electron (joules, J)

hf = incoming photon energy (joules, J)

hf_0 = energy required to eject an electron from a metal, also called the work function (joules, J)

f_0 = threshold frequency, i.e. the minimum frequency of a photon which causes electron emission

Rearranging this equation in terms of electron kinetic energy gives:

$$E_k = hf - hf_0$$

This is the equation of the straight-line graph shown in Figure 13.3.

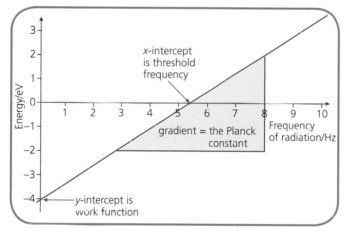

Figure 13.3 Graph of the photoelectric effect

So the gradient of the graph is Planck's constant, h, and the y-intercept gives the energy required to eject an electron from the metal surface, also called the work function.

This explained all of the experimental observations but, at the time, was a huge leap. Light was known to be a wave as it displayed interference patterns.

Figure 13.4 CCD (charged-coupled device) sensors utilise the photoelectric effect. They are used extensively in digital imaging

Einstein was proposing that on the atomic scale, light behaved as a particle. In fact, he proposed that light is both a particle and a wave at the same time: on the large scale it interferes and acts as a wave; on the small scale it acts as a particle and collides and imparts kinetic energy via those collisions.

Irradiance and the inverse square law

The energy associated with a beam of light depends on a number of factors. It is common for this to be referred to as the intensity of the beam but in physics, the term intensity has a very specific meaning which requires the measurement of the profile of the beam (i.e. its cross-section). A simpler concept (and the one we will use) is that of **irradiance**. Irradiance is defined using the following equation:

$$I = P/A$$

Table 13.1 details the variables of this equation.

This means that for the same power of light, a beam with a smaller cross-sectional area will have a larger irradiance.

Note that even with small power ratings (5 mW) the irradiance is a large value as the laser beam is very small in diameter. In addition, note that the irradiance

Quantity	Name	Unit	Symbol
I	irradiance	watts per square metre	$W\,m^{-2}$
P	power	watt	W
A	area	square metre	m^{-2}

Table 13.1

Distance/m	1.09	1.00	0.91	0.81	0.71	0.61	0.51	0.46
Irradiance/W cm^{-2}	0.10	0.12	0.14	0.16	0.20	0.27	0.37	0.47

Table 13.2 Irradiance versus distance data

falls as the beam becomes more spread out the further from the source it is measured. A laser is a typical light source as it produces a highly directional, narrow beam which does not spread out very much even over very long distances.

Worked Example 13.1

Commercially available pointers emit visible light between 532 nm (green) and 700 nm (red) and have power outputs ranging from less than 1 mW to 5 mW. The cross-sectional area of the beam itself is typically 3–6 mm² at the aperture, spreading out to 40–80 mm² at 3 m. Calculate the maximum irradiance at the aperture and at a distance of 3 m.

At the aperture:

$I = P/A$
$= 5 \times 10^{-3}/3 \times 10^{-6}$ (as 3 mm² $= 3 \times 10^{-6}$ m²)
$= 1.67 \times 10^{3}$ W m^{-2}

At 3 m distance:

$I = P/A$
$= 5 \times 10^{-3}/40 \times 10^{-6}$ (as 40 mm² $= 40 \times 10^{-6}$ m²)
$= 0.125 \times 10^{3}$ W m^{-2}

A point light source is one which emits light equally in all directions in three dimensions. In contrast to the laser in the above example, the light energy at any particular distance from the source may be imagined as being spread out over the surface of a sphere centred on the source. The source itself is not necessarily a small object, but its size must be small when compared to the distance of the observer measuring the emitted light. (In fact, any source of radiation may be considered as a

point source when it meets this criterion.) In this way, any star may be considered a point source when viewed here on Earth.

In an experiment to measure the relationship between the irradiance of a lamp and distance, the data given in Table 13.2 were recorded.

Plotting these data gives us the graph shown in Figure 13.5.

The relationship is clearly not one of direct proportion.

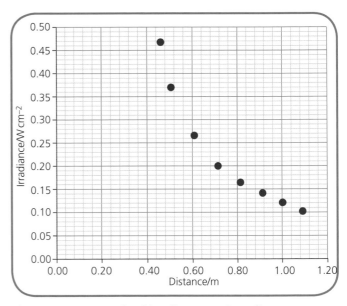

Figure 13.5 A graph of irradiance against distance

The irradiance versus distance relationship is actually given by the following equation:

$$I = k/d^2$$

Table 13.3 details the variables of this equation.

Quantity	Name	Unit	Symbol
I	irradiance	watt per square metre	W m^{-2}
d	distance	metre	m
k	constant of proportionality	watt	W

Table 13.3

This may be shown by plotting the graph of irradiance against $(1/(\text{distance})^2)$ as shown in Figure 13.6.

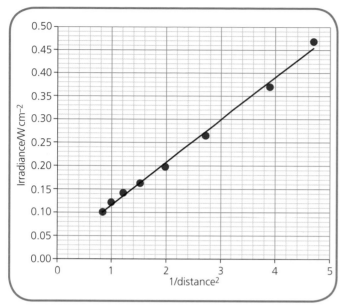

Figure 13.6 A graph of irradiance against $1/(\text{distance})^2$

Spectra

When a body is heated, in other words, when it becomes more energetic, it will begin to emit visible light. As early as the 1800s, physicists were aware that at any temperature a body will emit electromagnetic radiation and that the type of radiation emitted will depend on the temperature of the body itself. This is known as **blackbody radiation**. In the visible region, the spectrum of wavelengths emitted looks like the

white light spectrum produced by a prism and is a continuous range as shown in Figure 13.7.

In the early 1800s it was noted that the light coming from the Sun had a number of 'missing' wavelengths characterised by dark lines in the emission spectrum produced. These became known as Fraunhofer lines after the German physicist who first identified in excess of 500 of these wavelengths. (Chapter 6 includes a further discussion of Fraunhofer lines.)

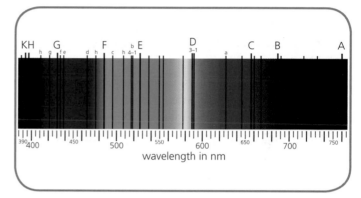

Figure 13.8 Fraunhofer lines

Around the same time, chemists like Bunsen (among others) were looking at the light emitted by chemical elements when heated in a flame in a laboratory. Passing this light through a spectrometer to analyse the spectrum produced showed that it was made up of a series of discrete lines. Each series was unique to a particular element and could be used as a type of fingerprint or coloured barcode to determine the element's presence. Figures 13.9 and 13.10 show the distinct patterns of lines of wavelengths for hydrogen and iron, respectively.

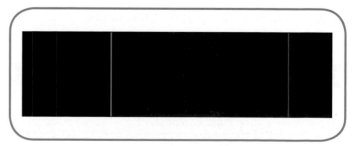

Figure 13.9 The visible spectrum for hydrogen

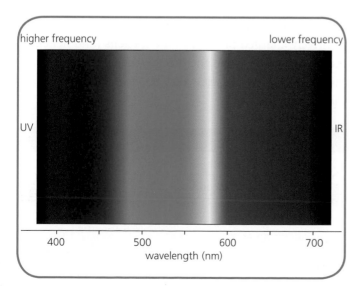

Figure 13.7 The visible spectrum

Figure 13.10 The spectrum for iron

It became apparent that the 'missing' lines in the Sun's continuous emission spectrum were matched with the emission lines of individual elements which could be measured here on Earth. As a result it was proposed that the white light being emitted by the surface of the hot Sun was passing through its atmosphere of chemical elements which were only absorbing the wavelengths that somehow characterised their structure. This discovery led to many further questions. Why do the wavelengths absorbed by hydrogen differ from those absorbed by iron? Why is it only these special wavelengths which are emitted or absorbed to the exclusion of all others? The answers to these questions had to wait a further 60 years or so and required Planck, Einstein and Bohr to pave the way.

Einstein developed Planck's work on blackbody radiation and proposed that, at the smallest level, light was delivered in 'packets' rather than as a continuous flow of energy. As noted previously, Einstein called these packets **photons** of light, proposing that they acted like light particles. Each particle carried a fixed amount of energy given by the equation:

$$E = hf$$

Table 13.4 details each part of this equation.

This means that each photon of light absorbed or emitted by an atom has a particular, unique energy. Bohr's model uses this fact and Rutherford's nuclear atomic model to propose the following:

- Electrons orbit a dense, positively charged nucleus in a manner similar to planets around a star.

- The radius of the orbit is determined by the electrical potential energy of the electron in the electric field created by the nucleus.

- As the electrons are attracted to the nucleus, work needs to be done to move them further away and so larger radius orbits are higher-energy states.

- Most significantly, only particular energy states are allowed. This means that only particular orbit radii are permitted.

- To jump from a low-energy state to a high-energy state requires the electron to absorb exactly the right amount of energy from an incoming photon. This is known as an electron excitation with the electron being promoted to a higher energy state.

- Conversely, if an electron falls from an excited state to a lower state it will emit a photon whose energy will correspond exactly to the energy gap in the two states.

The proposal to restrict the energy states to a limited set of discrete values explains the observed absorption and emission spectra. The characteristic lines or wavelengths emitted or absorbed (missing) correspond exactly to the energy gap in two electron states. As different elements contain different numbers of positively charged protons in the nucleus (and hence number of electrons in the orbits), the exact values for the energy states and their differences are unique to those elements.

Quantity	Name	Unit	Symbol
E	energy of photon	joule	J
h	Planck's constant	joule per second	$J\,s^{-1}$
f	frequency of light	hertz	Hz

Table 13.4 Einstein-Planck equation variables

Quantity	Name	Unit	Symbol
hf	energy of photon emitted or absorbed	joule	J
E_1	electron energy in initial state	joule	J
E_2	electron energy in final state	joule	J

Table 13.5

By conservation of energy, this interaction may be represented by the following equation:

$$E_2 - E_1 = hf$$

Table 13.5 details each part of this equation.

The electron energy levels in an atom may then be represented as shown in Figure 13.11.

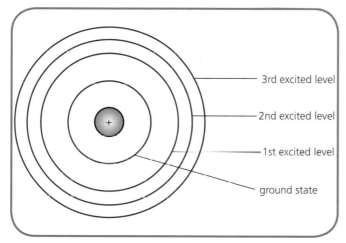

3rd excited level

2nd excited level

1st excited level

ground state

Figure 13.11 The lowest four 'allowed' energy levels for an atom of hydrogen

Note that in the absence of external energy input, electrons will tend to occupy the lowest possible energy levels. This arrangement is known as the **ground state**. For a hydrogen atom with a single proton nucleus and a single orbiting electron, the electron would orbit at the closest 'allowed' radius. The ground state energy is often referred to as E_1.

If an atom is ionised, in other words, loses an electron completely, then that electron is assumed to be at infinite radius. At this separation, the electrical potential energy of the electron in the field of the nucleus is zero. Given this definition, the absolute value of all other energy levels is defined as negative. Differences between energy levels are still calculated by subtracting final from initial value ($E_2 - E_1$) and may be positive where energy is 'added' to the atom by absorption of a photon or negative where energy is 'removed' from the atom by emission of a photon.

The Bohr model only accurately predicts the line spectra for hydrogen-like atoms – that is atoms where there is only one electron orbiting the nucleus, such as H or He^+. Nevertheless it was a huge leap in the understanding of atomic structure and provided a method of 'fingerprinting' elements.

For a hydrogen atom, the energy levels are determined by the use of a quantum number, n, where n is a positive integer used to describe the possible energy level occupied by the electron. The Bohr model was one of the first applications of a quantum approach where the allowed values of a variable (in this case electron energy) could only take one of a set of particular values. The ground state energy value, E_1, may then be used to calculate the energy of any other state using the formula:

$$E_n = E_1/n^2$$

Table 13.6 details each part of this equation.

Quantity	Name	Unit	Symbol
n	quantum number describing n^{th} energy level	none	none
E_1	ground state energy value	joule	J
E_n	electron energy in n^{th} state	joule	J

Table 13.6

The ground state energy may be measured in joules, but it is also common to see the value in electron volts (eV), where $1\text{eV} = 1.602 \times 10^{-19}$ J. Table 13.7 shows the energy values for the first few energy levels of the hydrogen atom calculated using the formula above.

n	Energy/eV	Energy/J
6	−0.38	-6.05×10^{-20}
5	−0.54	-8.71×10^{-20}
4	−0.85	-1.36×10^{-19}
3	−1.51	-2.42×10^{-19}
2	−3.40	-5.45×10^{-19}
1	−13.61	-2.18×10^{-18}

Table 13.7 Atomic hydrogen electron energy levels

Given the data in the table, it is then possible to predict the frequency (and wavelength) of the photon associated with any transition. For example, the energy difference between levels 3 and 2 is given by

$$E_3 - E_2 = -2.42 \times 10^{-19} - (-5.45 \times 10^{-19})$$
$$= 3.03 \times 10^{-19}\,\text{J}$$

For the photon,

$$E = hf$$
$$3.03 \times 10^{-19} = 6.63 \times 10^{-34} \times f$$
$$f = 3.03 \times 10^{-19}/6.63 \times 10^{-34}$$
$$= 4.57 \times 10^{14}\,\text{Hz}$$

and so, $\lambda = 6.56 \times 10^{-7}$ m (as $v = f\lambda$)

This corresponds to the red wavelength line measured experimentally in the hydrogen line spectrum. (This wavelength is one of the 'Balmer series' of lines.)

Consolidation Questions

1 Given that the equation for the photoelectric effect may be written as $E_k = hf - hf_0$:

   a) Write down the meanings and units of the symbols in the equation.

   b) Sketch a graph of the ejected electron kinetic energy versus incident photon frequency for a photoelectric effect experiment. (Remember graph sketches need axes labels and units along with any numbers you can include.)

   c) Explain how might this graph be used to determine Planck's constant?

   d) Add a new line to your graph to show the results if the metal is exchanged for one with a greater work function.

2 Light of wavelength 454 nm is incident on a metal surface and causes electrons to be emitted with a maximum kinetic energy of 3.36×10^{-19} J.

   a) Explain why this light causes the emission of electrons.

   b) Find the work function for this metal.

   c) What is the maximum kinetic energy of the electrons if the light source power is doubled?

3 State three features of the photoelectric effect experimental result that cannot be explained by a wave-based theory of light.

4 Given that the irradiance of the Sun is estimated as $6 \times 10^7\,\text{W m}^{-2}$ at its surface, what is the irradiance of the Sun's energy incident on the Earth, some 1.5×10^{11} m distant? (Radius of the Sun is 6.95×10^8 m.)

5 The irradiance, I, from a point source is measured at various distances, r, from the source.

   a) Sketch a graph of I as a function of distance, r. (Remember graph sketches need axes labels and units along with any numbers you can include.)

   b) What must be done to the data in order to plot a graph of direct proportion between the experimental variables?

6 The Balmer series is made up of a number of frequencies in the hydrogen line spectrum. Complete Table 13.8 showing the wavelength, frequency or energy of each line.

Wavelength/nm	Frequency/Hz	Energy/J
656.4	4.57×10^{14}	3.03×10^{-19}
		4.09×10^{-19}
434.1		
	7.31×10^{14}	

Table 13.8

7 Using the data in Figure 13.12, answer the following questions:

a) Calculate the ionisation energy for a hydrogen atom (the energy that must be supplied to remove the electron from the atom entirely).

b) What wavelength would be emitted when an excited electron falls from the $n = 4$ to the $n = 2$ level? What colour light would this be?

c) For the blue line of wavelength 434.1 nm, which energy levels are involved in the electron transition?

d) Explain why it is not possible to see the radiation emitted when an electron returns to the ground state (lowest energy level).

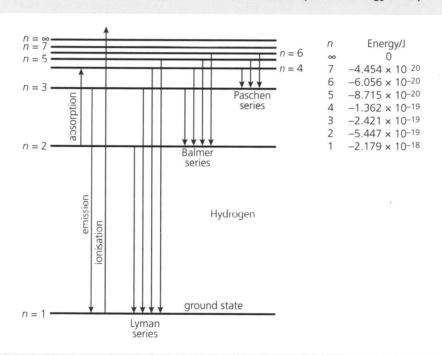

Figure 13.12

3

Electricity

14 Electrons and energy

Monitoring and measuring AC

AC and DC

Most of us have heard the terms AC and DC, but we may not know what they mean. AC stands for **alternating current** and DC for **direct current**. To be able to understand what these terms mean, we need to look at what current is.

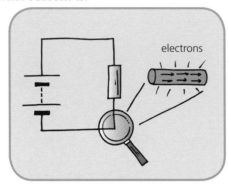

Figure 14.1 Electrons flow round a DC circuit

Current can be described as the flow of charge measured in amperes (A). One ampere represents one coulomb of charge passing a point in one second. In circuits, the charge carrier is usually the electron. The charge on one electron is approximately $-1.60 \times 10^{-19}\,\text{C}$ (or $-0.000\,000\,000\,000\,000\,000\,160\,\text{C}$) which means that for a current of 1 A there must be almost 6.25×10^{18} (or $6\,250\,000\,000\,000\,000\,000$) electrons passing a point in one second. While this seems like an enormous number, electrons are incredibly small (less than 10 femtometres) and so it is not unusual to have this size of current in a circuit.

Since we know that current is the amount of charge per second, we can create an equation relating charge, current and time:

$$I = \frac{Q}{t}$$

where

I is the current, measured in amperes (A)
Q is the charge in coulombs (C)
t is the time measured in seconds (s).

Normally this will be rearranged to give us

$$Q = It$$

Worked Example 14.1

1 How much charge has passed a point in a circuit after 5 s if there is a current of 0.5 A?

$Q = ?$
$t = 5\,\text{s}$
$I = 0.5\,\text{A}$
$Q = It = 0.5 \times 5 = 2.5\,\text{C}$

The charge is 2.5 C.

2 After 20 s, 8 C of charge has passed a point in a wire. What is the average current in the wire?

$Q = 8\,\text{C}$
$t = 20\,\text{s}$
$I = ?$
$Q = It$
$8 = I \times 20$
$I = 8/20 = 0.4\,\text{A}$

The average current in the wire is 0.4 A.

It becomes more complex when we try to explain in detail what is happening to the electrons in a circuit. In direct current this is relatively simple: the electrons move along the circuit in one general direction at a very slow rate.

We relate the movement of the electrons to their average velocity, because individual electrons are constantly bouncing off one another and do not flow in perfectly straight lines along the circuit as we may imagine. The velocity of the electrons will depend on the material (and as a result how many 'free' electrons there are), the area through which they can flow and the current in the wire. In a 1 mm^2 copper wire with a 1 A current in it, we can expect an average velocity in the region of 0.1 mm s^{-1}. In a DC circuit the average velocity of the current is always in one direction.

When it was first being investigated, before the discovery of the electron, electricity was assumed to be the flow of positive particles from one point to another. Electrons are negatively charged, however, and as a result current is measured in the opposite direction to the movement of electrons. In a DC circuit, electrons will move from the negative terminal towards the positive terminal, but current (in many texts) is measured as positive if it moves from the positive terminal to the negative one.

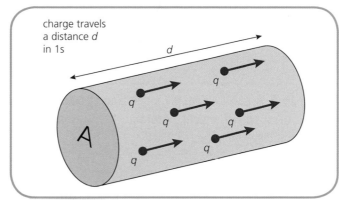

Figure 14.2 Simple copper wire with a current flowing through it. The individual electrons move very slowly along the wire

In an alternating current circuit the movement of the electrons is less obvious. Alternating current is so named because the current is constantly changing direction. This means that the electrons inside the circuit must go one way then the other, constantly changing at double the rate of the frequency of the current. As a result, electrons do not actually travel but simply vibrate within a very small area.

How is energy transferred along a wire if the electrons move so slowly in DC and only vibrate in an AC system? The answer is that the electric and magnetic fields (detailed more thoroughly in Chapter 9) actually carry the signals and most of the energy. While the movement of electrons carries a small amount of energy, the fields around the wire are the defining factor in the transfer of energy.

Research Task

Why is the potential difference and frequency of the UK mains electricity system what it is? Is there a historical reason or does it lie in reliability?

What systems do other countries use and why?

Peak and root mean square

The measurement of a DC potential difference is clear – it is simply the value which is measured or, for example, the reading on a supply. This is not the case for AC potential differences. The value which we speak about is the effective or equivalent value. It is the **root mean square** value or V_{rms} as it will be written from now on. This value is actually lower than the peak value which we will get if we look at the trace.

Figure 14.4 on the next page shows the waveform of the UK mains system. The peak voltage is measured from the centre of the waveform to the top. This is approximately 325 V for the UK system but as you may know, the 'rated' value is 230 V. The relationship is to do with the shape of the waveform and it works out as:

$$V_{peak} = \sqrt{2}\,V_{rms}$$

| **For Interest** | **The war of currents** |

In the late 1800s a battle raged between two competing camps. Thomas Edison and his company General Electric owned some very important patents for the running of a DC network. George Westinghouse owned a company which held similar rights to AC distribution systems. AC was championed on behalf of Westinghouse by Nikola Tesla, the designer and inventor of important AC generators, who had previously been employed by Edison. Very public campaigns were put in place to try to discredit the other electricity style, including suggestions that one was safer than the other.

Ultimately AC was adopted almost universally because it was more efficient in transferring electrical energy over long distances. Edison found it very difficult to give up, though, and as late as 1902 he demonstrated the 'dangers of AC' by electrocuting animals in public. Today most power networks work with AC because the efficiency is much greater, though in our homes many appliances, such as computers, require direct current.

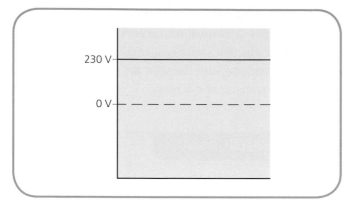

Figure 14.3 The output of a 230 V DC supply – a simple, straight line

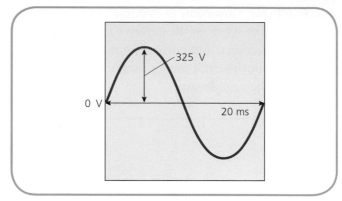

Figure 14.4 A 230 V AC supply, operating at 50 Hz, just like the UK supply. The peak potential, 325 V, is much higher than the rms value of 230 V

Similarly, we can calculate the peak and rms currents:

$$I_{peak} = \sqrt{2}\, I_{rms}$$

We can also measure the frequency of AC waveforms by working out how much time there is between peaks. To do this we use an oscilloscope.

For Interest	Potential difference or voltage?

We have investigated what current is but we have not yet defined what we mean by a voltage. Voltage is often referred to as potential difference or p.d., and this term gives us a clue as to what voltage really is. Every point on a circuit has an electrical potential and the difference between the electrical potential at two points is therefore referred to as the potential difference. The volt is a measure of energy (or work done) per unit charge; in other words, 1 V is equal to 1 joule for every coulomb of charge. This is shown by the equation $V = E_w/Q$.

The oscilloscope

An **oscilloscope** is an instrument used in the investigation of electrical signals. The screen displays a trace of the electrical signal and in almost all cases this is a graph of potential difference versus time.

We can measure the potential of a DC trace very simply as the number of vertical squares, or divisions, multiplied by the number of volts per square. Similarly, we can measure the peak AC potential difference by first measuring the number of divisions from the middle of the trace to the top. It is essential to confirm what the scale is and we are able to vary this using the dials on the front of the oscilloscope. The time is measured in the same way, using the horizontal divisions, and this allows us to measure both the frequency and period of the signal.

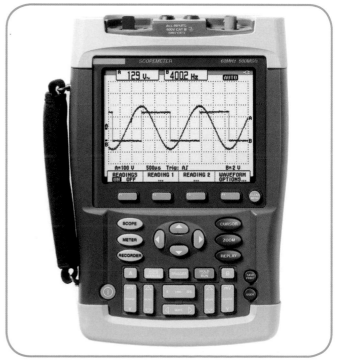

Figure 14.5 A digital oscilloscope

Worked Example 14.2

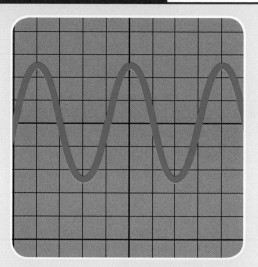

Figure 14.6

Find the following attributes of the signal shown in Figure 14.6 if the oscilloscope is set to measure 5 V per vertical division and 10 ms per horizontal division:

a) V_{peak}

b) V_{rms}

c) period of the wave

d) frequency of the wave.

a) The peak of the voltage is the distance from the centre point to the top of a peak and the trace covers 2.5 divisions, so $V_{peak} = 2.5 \times 5 = 12.5\,V$.

b) The rms value is measured as $\dfrac{V_{peak}}{\sqrt{2}}$ which gives us 8.8 V.

c) The wave, from peak to peak, takes four horizontal divisions, so $t = 4 \times 0.01 = 0.04\,s$.

d) The frequency of a wave is $1/t = 25\,Hz$.

Questions

1 A kettle takes 2 minutes to boil some water. During that time it draws a current of 6 A. How much charge passes through the kettle?

2 How many electrons must pass through a wire if it carries a 5 A DC current for 10 seconds?

3 Describe the movement of electrons inside wires carrying

 a) a direct current

 b) an alternating current.

4 In Figure 14.7, electrons are shown moving in a wire carrying a direct current. Redraw the circuit showing the direction of the current.

5 The electricity system in America uses a different potential difference to that in Scotland. Appliances in the US state that they are rated at 110 V. A student in New York checks this using an oscilloscope and sees a much larger potential difference.

 a) What is the value the student sees?

 b) Why is this different from the 'rated' value?

Figure 14.7

6 Look at the oscilloscope trace (Figure 14.8). What is:

 a) the peak voltage

 b) the rms voltage

 c) the frequency of the signal?

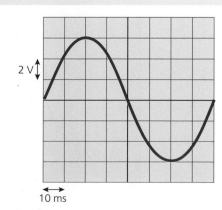

2 V

10 ms

Figure 14.8

7 Two circuits are set up side by side as shown in
   Figure 14.9. Which lamp will be brighter?

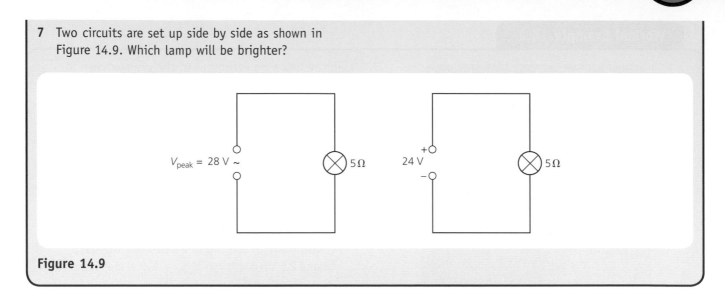

Figure 14.9

display

power button

DC ammeter

AC ammeter

ammeter input

low current
ammeter input

ohmmeter

DC voltmeter

AC voltmeter

voltmeter and
ohmmeter input

common input

Figure 14.10

Using a multimeter

Multimeters are widely used in engineering and physics when we investigate circuits. They are particularly useful because they allow us to investigate multiple properties of circuits with only one device. They can be used instead of separate voltmeters, ohmmeters and ammeters.

Figure 14.10 shows a standard multimeter, but the vast array of options can make them intimidating. We will look at how the multimeter dial settings should be used to change the meter's use. In all cases, the range of numbers tells us the maximum reading the meter could give on that setting. For example, the 20 k mark on the ohmmeter section means that the maximum you can investigate is 20 kΩ. If we had 25 kΩ, we would have to increase the meter to the 200 kΩ setting. You may think that it would always be best to select the highest reading available, but this may reduce the accuracy of the measurement. It is best to start with the highest value and gradually reduce the setting.

The multimeter as a DC voltmeter

1 Connect one cable to the common input and one to the voltmeter input.

2 Select the highest DC voltmeter range possible (unless you know roughly where the reading should be, in which case you choose the most suitable) and attach the probes across the points you want to measure.

3 You can make the reading more accurate by reducing the range of the meter.

4 Once you have found the most accurate reading you can note down your measurement from the display.

To use the multimeter for AC, follow the instructions for the DC voltmeter, but select the highest AC voltmeter range in step 2.

Note: Be VERY careful when doing this – a simple multimeter should NOT be used to investigate mains electricity.

The multimeter as an ohmmeter

1 Connect one cable to the common input and one to the ohmmeter input.

2 Select the highest resistance range possible (unless you know roughly where the reading should be, in which case you choose the most suitable) and attach the probes across the resistor.

3 If you can make the reading more accurate, reduce the range of the meter.

4 Once you have found the most accurate reading you can note down your measurement from the display.

The multimeter as a DC ammeter

1 This is a little different. Connect one cable to the common input and one to the ammeter input.

2 Select the 10 A DC current range and attach the probes to your circuit, in series, at the point you want to measure current.

3 If the reading is very very small (less than 0.2 A), you may wish to use a more accurate reading. In this case, you need to move from the ammeter input to the low current ammeter input and change from 10 A to 200 mA in the ammeter range. If the current is larger than 200 mA, the fuse in the ammeter will be blown and it will no longer work.

Electrons and energy

Ohm's Law

You should be familiar with Ohm's Law from earlier years of study. This simple equation is the basis of many circuits where we attempt to explain the relationship between potential difference, current and resistance. This allows us to design and produce circuits which are capable of performing the tasks we want, be this to turn on a light or to find an elementary particle in the Large Hadron Collider (LHC).

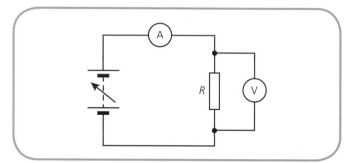

Figure 14.11

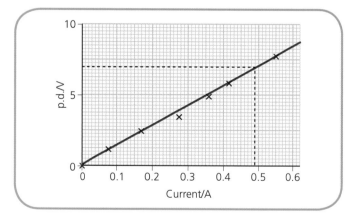

Figure 14.12 Graph of voltage against current

It is worthwhile to investigate Ohm's Law. Connect the circuit shown in Figure 14.11 and look at the meter readings when you vary the p.d. You should note that the current increases as the p.d. increases. If you repeated the experiment using different values of resistance, you would find that the current decreases as the resistance increases.

The graph in Figure 14.12 shows the sort of relationship you would obtain if you performed the experiment. We can see that as p.d. increases, current increases. V is proportional to I.

As we can see, $V \propto I$

So $V = K \times I$

Therefore $\frac{V}{I} = K$

And this constant, K, is known as resistance. Thus

$$\frac{V}{I} = R$$

$$V = IR$$

The circuit shown in Figure 14.11 will not do anything useful – the resistor will just get a little warmer.

For Interest **Georg Ohm**

Georg Ohm was a German mathematician who left university with a doctorate in mathematics and became a mathematics teacher. After moving to a different school, he was required to teach physics as well as mathematics and became interested in experimenting. This led him to discover a proportional relationship between current and p.d. As a result of this work he decided to publish a book in 1827 which includes what we now know as Ohm's Law. After his death, the unit of resistance was designated as the ohm in his honour.

For Interest **Resistance**

Resistance is a measure of how 'difficult' it is for electrical current to flow. The greater the resistance of a material, the more difficult it is for charge to pass through the material. When we refer to a resistor we are talking about a component which is placed in a circuit to reduce the current. Current will still pass through the resistor but there will be a drop in p.d. across the resistor, because potential – some energy – was 'used' in passing charge through the resistor.

It may seem odd that we would want to make it more difficult for current to flow but resistors are used to ensure that the p.d. across certain components of a circuit is what is required by those components. Every component has resistance so a potential drop occurs if we attach a lamp to a circuit instead of a resistor. Even connecting wires have a very small resistance so in really precise electronic devices, even the resistance of the conducting metals has an effect.

Figure 14.13 shows a few of the components you will need to use when designing more advanced or more practical circuits.

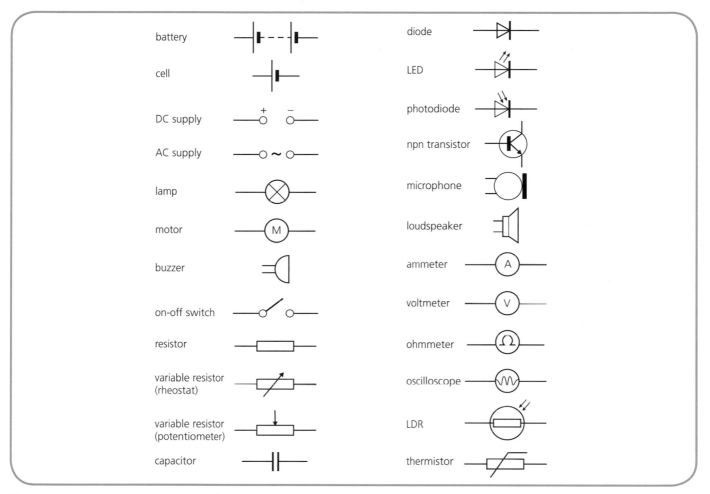

battery		diode	
cell		LED	
DC supply		photodiode	
AC supply		npn transistor	
lamp		microphone	
motor		loudspeaker	
buzzer		ammeter	
on-off switch		voltmeter	
resistor		ohmmeter	
variable resistor (rheostat)		oscilloscope	
variable resistor (potentiometer)		LDR	
capacitor		thermistor	

Figure 14.13 Common circuit symbols

Worked Example 14.3

Calculate the resistance of the lamp if the ammeter reads 1.5 A.

$V = 12\,V \qquad I = 1.5\,A$

$R = ?$

$V = IR$

$12 = 1.5 \times R$

$R = 12/1.5 = 8\,\Omega$

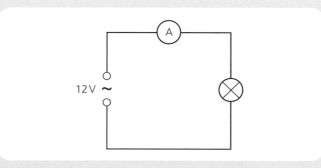

Figure 14.14

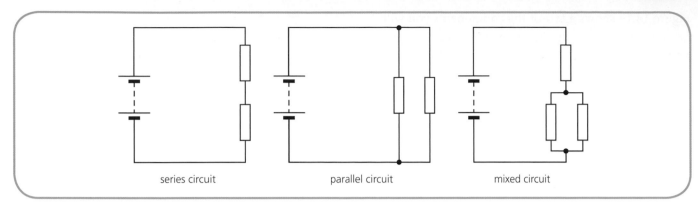

Figure 14.15 Circuit combinations

Series and parallel

We classify circuits in terms of whether they are in **series** or in **parallel**. This simply describes how the components are placed in the circuit. It is important to note that it is possible to have series and parallel sections in the same circuit.

In theory, all the circuits in Figure 14.15 can be reduced to a potential source and a single resistor. To do this, we need to know how to combine resistors. It is useful at this point to consider how current flows. We noted earlier that current flows through the resistor and that the resistor makes it more difficult to flow and 'uses up' some of the potential.

We will now consider three types of circuit – series circuits, parallel circuits and mixed circuits – and investigate their properties.

Series circuits

If we put two or more resistors together in series (in a row), it is the same as having one larger resistor on its own – it will oppose the flow of current in exactly the same way.

The voltage across the supply equals the sum of the voltages across each of the resistors.

$$V_T = V_1 + V_2 + \ldots$$

So $$IR_T = IR_1 + IR_2 + \ldots$$

As the current is the same in a series circuit, we can remove I to get

$$R_T = R_1 + R_2 + \ldots$$

where the dots indicate this continues for however many resistances there are.

Worked Example 14.4

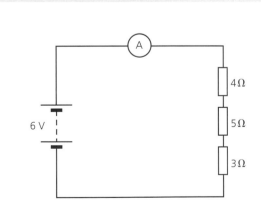

For the circuit shown in Figure 14.16, calculate the total resistance and the reading on the ammeter.

First, we calculate the resistance – there are three resistors, so we use the equation

$R_T = R_1 + R_2 + R_3$ which gives us

$R_T = 4 + 5 + 3 = 12\,\Omega$

The next stage is to redraw the circuit. This allows us to understand what the circuit looks like now (Figure 14.17).

Figure 14.16

We can now calculate the reading on the ammeter.

$V = 6\,V$

$I = ?$

$R = 12\,\Omega$

$V = IR$

$6 = I \times 12$

$I = 6/12 = 0.5\,A$

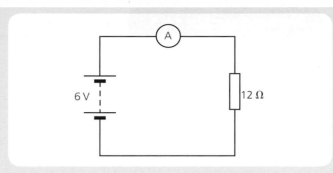

Figure 14.17

This leads on to some important features of series circuits. The current in a series circuit is the same at all points in the circuit. In our example above, each of the three resistors will have the same current through it. This has implications for the p.d. across each resistor because $V = IR$, so V across each resistor depends on the resistance.

There are a number of principles that apply in series circuits:

1 The current is the same at all points in the circuit.

2 The resistances can be added together to work out the total resistance.

3 The p.d. across each resistor depends on the resistance.

4 The total p.d. is equal to the total of the potential differences across the individual resistors.

Parallel circuits

If we place two resistances alongside each other, we create two paths for charge to flow through. This results in less charge trying to flow through each of the resistances. Essentially this makes it easier for current to flow.

We can use a comparison with a tunnel to help us understand. If there is only one tunnel, only a certain amount of charge can push through the tunnel at one time. If we have two tunnels, however, the charge will split between the two tunnels and the total passing through the tunnels is greater.

This is similar to what happens in a parallel circuit. The current splits and goes along each of the available paths. This means that while each current on a branch of the parallel circuit can be different, the total current must add up to the total current coming from the supply.

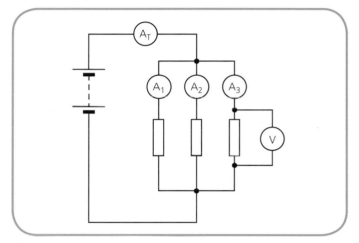

Figure 14.18 Parallel circuit: the ammeter reading on A_T must equal $A_1 + A_2 + A_3$, but each resistor has the same p.d. across it

We can use this fact to deduce a formula for the total resistance in a parallel circuit. The total current (I_T) equals the current through each resistor, so

$$I_T = I_1 + I_2 + I_3$$

From Ohm's Law $I = V/R$ so

$$\frac{V}{R_T} = \frac{V_1}{R_1} + \frac{V_2}{R_2} + \frac{V_3}{R_3}$$

In the circuit shown in Figure 14.18, the potential difference across each resistor is the same as the supply.

This means that $\quad V = V_1 = V_2 = V_3$ so

dividing $\quad \dfrac{V}{R_T} = \dfrac{V_1}{R_1} + \dfrac{V_2}{R_2} + \dfrac{V_3}{R_3}$ by V

gives us $\quad \dfrac{1}{R_T} = \dfrac{1}{R_1} + \dfrac{1}{R_2} + \dfrac{1}{R_3}$

Similar to the equation for series resistors, this can be extended for any number of resistors in parallel.

Worked Example 14.5

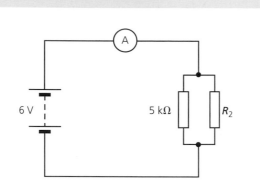

Figure 14.19

The circuit in Figure 14.19 has a total resistance of $R_T = 3.3\,k\Omega$. Calculate the value of R_2 and the current in this resistor.

The first step is to work out the resistance.

$R_T = 3.3\,k\Omega$
$R_1 = 5\,k\Omega$
$R_2 = ?$

$$\frac{1}{R_T} = \frac{1}{R_1} + \frac{1}{R_2}$$

$1/3300 = 1/5000 + 1/R_2$

$1/R_2 = 1/3300 - 1/5000 = 0.0003 - 0.0002 = 0.0001$
$R_2 = 1/0.0001 = 10\,000 = 10\,k\Omega$

We can then calculate the current using Ohm's Law

$R = 10\,k\Omega$
$V = 6\,V$
$I = ?$
$V = IR$
$6 = I \times 10\,000$
$I = 6/10\,000 = 0.0006\,A = 0.6\,mA$

The principles we need to remember about parallel circuits are:

1 The resistances do not simply add together; we need to use the equation to calculate the combined resistance.

2 The current in each of the branches of the parallel circuit adds up to the total current in the circuit.

3 The current in each branch depends on the resistance: the lower the resistance, the higher the current.

4 The p.d. across each branch of the parallel circuit is the same.

Mixed circuits

At first, mixed circuits may seem complicated, but if we separate them into smaller chunks, they become much simpler.

Take the circuit in Figure 14.20, for example. We want to know the total resistance of the circuit. The first step is to calculate the series resistance on the right-hand parallel branch.

There are two resistors in series on one branch. In a case like this we calculate the total resistance in that branch first. The series resistance equation is $R_T = R_1 + R_2$, so in the branch the resistance is $200\,\Omega + 300\,\Omega$ which gives us $500\,\Omega$.

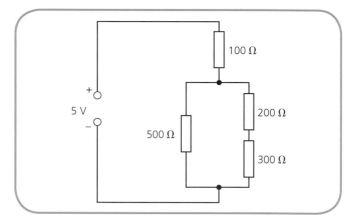

Figure 14.20

The next stage is to redraw the circuit, as shown in Figure 14.21.

We can see that the two resistors have now been combined leaving a two-branch parallel circuit.

We apply the parallel resistance equation

$\frac{1}{R_T} = \frac{1}{R_1} + \frac{1}{R_2}$ which in this case gives us

$$\frac{1}{R_T} = \frac{1}{500} + \frac{1}{500}$$

$$\frac{1}{R_T} = 0.002 + 0.002 = 0.004$$

So the total parallel resistance is 1/0.004 which is $250\,\Omega$.

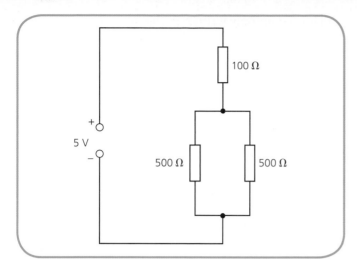

Figure 14.21

We then redraw the circuit again, as shown in Figure 14.22.

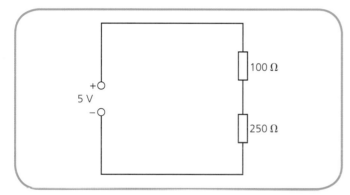

Figure 14.22

This is now a series circuit. We combine the two remaining resistances together to give us our final total resistance: $100\,\Omega + 250\,\Omega = 350\,\Omega$.

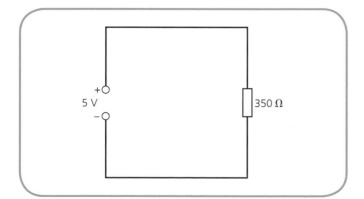

Figure 14.23

We have reduced a complex circuit into a simple, recognisable one. At any stage in this process you could use Ohm's Law to calculate various currents

and potential differences. It is important to recognise whether the section of circuit you are working on is in series or parallel, because you will meet mixed circuits like this.

Potential dividers

A **potential divider** is the name we give to a circuit which uses components to split up the supply voltage. Sometimes it is necessary to supply a circuit with one potential, but have another section requiring a lower p.d. When this happens, we can use resistors to divide the current.

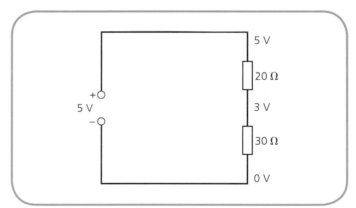

Figure 14.24

In Figure 14.24 we can see that the supply is split between the two resistors. The total current through each resistor must be the same – in this case we can calculate it using Ohm's Law to be 0.1 A because the total resistance is $50\,\Omega$.

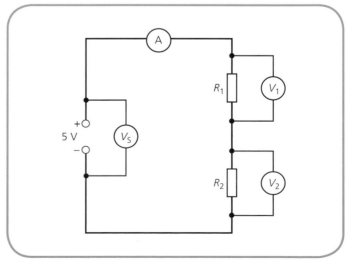

Figure 14.25

If we then use the same equation for each resistor, we find that the potential across the $20\,\Omega$ resistor is $0.1 \times 20 = 2\,V$ and that across the $30\,\Omega$ resistor is $0.1 \times 30 = 3\,V$.

We can work out a general rule for calculating the potential across each resistor.

We know that the total resistance in the circuit in Figure 14.25 is $R_1 + R_2$. From Ohm's Law we know that $V_S = IR_T$, where V_S stands for V_{supply}, so we can substitute R_T to give $V_S = I(R_1 + R_2)$.

Rearranging this for current we find that
$$I = \frac{V_S}{(R_1 + R_2)}$$
We also know that the current in each of the resistors is the same as the current throughout the circuit, so
$$I = \frac{V_1}{R_1} = \frac{V_2}{R_2}$$
As we know each of the currents is the same we can say
$$\frac{V_1}{R_1} = \frac{V_S}{(R_1 + R_2)}$$
(both of these equations give us the current).

Rearranging this equation, we can calculate V_1:
$$V_1 = \frac{R_1 V_S}{(R_1 + R_2)}$$

which is commonly written in the form
$$V_1 = \frac{R_1}{(R_1 + R_2)} \times V_S$$
We can derive a similar equation for V_2
$$V_2 = \frac{R_2}{(R_1 + R_2)} \times V_S$$
Additionally, if we already know some of the details of the circuit, we can use the relationship $I = \frac{V_1}{R_1} = \frac{V_2}{R_2}$ rearranged in the form $\frac{R_1}{R_2} = \frac{V_1}{V_2}$, which shows that the ratio of the potentials is equal to the ratio of the resistances.

Potential dividers may be used as 'voltage controllers'. This means that they are used to control the amount of potential across two terminals. Often we will use **potentiometers** and **rheostats** in circuits like this. Potentiometers and rheostats are two types of variable resistor.

Figure 14.26 A potentiometer

Worked Example 14.6

Calculate V_1 and V_2 in the circuit given in Figure 14.27.

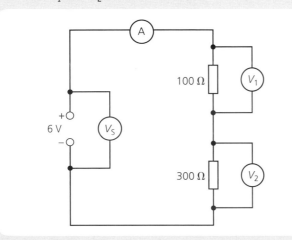

Figure 14.27

$V_S = 6\,V$
$V_1 = ?$
$V_2 = ?$
$R_1 = 100\,\Omega$
$R_2 = 300\,\Omega$
$R_T = 100 + 300 = 400\,\Omega$
$V_1 = [R_1/(R_1 + R_2)]V_S$
$V_1 = [100/(100 + 300)] \times 6$
$V_1 = 100/400 \times 6 = 0.25 \times 6$
$V_1 = 1.5\,V$

There are a few ways to work out V_2 from this point on. We could calculate it using the equivalent equation, we could use the $R_1/R_2 = V_1/V_2$ relationship, or we could use the simple rule from series circuits – the total potential is the sum of the individual potentials.

So $V_S = V_1 + V_2$ and as a result
$V_2 = V_S - V_1 = 6 - 1.5 = 4.5\,V$.

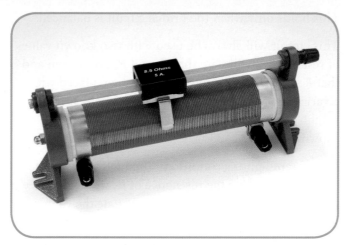

Figure 14.28 A rheostat

Potentiometers or 'pots' are more commonly found in simple potential dividers. This is because the three terminals on a pot allow it to be used as two separate resistors – the middle connection is like the wire connecting the two resistors. If you turn the pot one way, R_1 gets higher and R_2 gets lower; if you turn it the other way, the opposite happens. This means that you are controlling two different resistance values with one operation.

You will find the sort of circuit in Figures 14.29 and 14.30 at home in volume controls or 'dimmer switches'. Potential dividers could also be used with other circuits. They are often used with **transistors** which act a little like switches as shown in Figure 14.31. They allow current to flow through them when they are switched on – so when the transistor is on in this circuit, so is the LED.

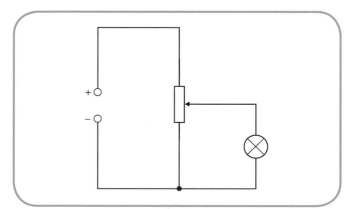

Figure 14.29 The potentiometer acts like two resistors: turning the potentiometer will change both values at once. This would work as a dimmer switch, because the potential across the lamp would be altered depending on the resistances on the potentiometer

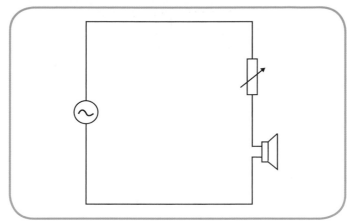

Figure 14.30 This circuit uses the rheostat as a volume control. The potential difference of the signal generator's output is shared between the rheostat and the loudspeaker. If the resistance of the rheostat is reduced, the volume goes up because the loudspeaker's 'share' of the potential increases

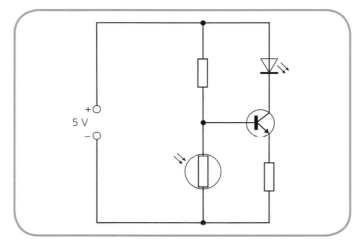

Figure 14.31 A simple night light circuit. The LED will come on when the transistor switches on. As the light decreases, the resistance of the LDR increases, as does its 'share' of the potential, turning the transistor on

If we wanted to be able to adjust the light level required to turn this circuit on, which components could we change? How could we alter it to be a temperature-dependent circuit? Try out some designs and if the equipment is available, build the circuits and investigate how they work.

Wheatstone bridges

Potential dividers can be used in a 'bridge' formation, as shown in Figure 14.32 on the next page. The voltmeter measures the potential difference between the two midpoints of the branches, allowing us to find out whether the circuit is balanced. If the reading is 0 V, we say the circuit is balanced.

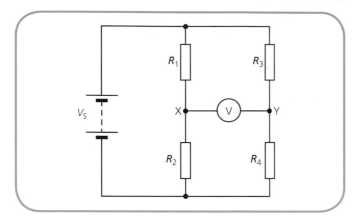

Figure 14.32 Potential divider in a 'bridge' setup. The voltmeter measures the potential difference between the midpoints of the two branches

We know that the potential at X will be the potential across R_2, so we can use the traditional potential divider formula:
$$V_2 = \frac{R_2}{(R_1 + R_2)} \times V_S$$

Similarly, the potential at Y will be the potential across R_4:
$$V_4 = \frac{R_4}{(R_3 + R_4)} \times V_S$$

If the voltmeter reads 0, then V_2 and V_4 must be equal, so:
$$V_4 = V_2$$
$$\frac{R_2}{(R_1 + R_2)} \times V_S = \frac{R_4}{(R_3 + R_4)} \times V_S$$

Dividing by V_S this becomes
$$\frac{R_2}{(R_1 + R_2)} = \frac{R_4}{(R_3 + R_4)}$$

By cross-multiplying the fractions, we get
$$R_1R_4 + R_2R_4 = R_2R_3 + R_2R_4$$

Subtracting R_2R_4 from both sides leaves $R_1R_4 = R_2R_3$ which we can rearrange to arrive at the ratio:
$$\frac{R_1}{R_2} = \frac{R_3}{R_4}$$

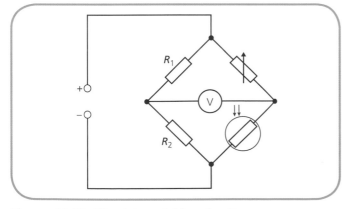

Figure 14.33 Wheatstone bridge circuit

This holds only when the bridge circuit is balanced.

This circuit will always be used with two known values of resistor. We will state that R_1 and R_2 are chosen and that R_3 and R_4 can change. R_3 might be replaced with a variable resistor and R_4 with a component such as a light-dependent resistor (LDR) to give the circuit shown in Figure 14.33.

In different lighting conditions, the LDR's resistance will change. By altering the resistance of R_3 we can balance the circuit, which means we can calculate the value of the resistance of the LDR. This allows us to use the LDR as a light meter, as we will know what resistance the LDR should have for a given light intensity. This set-up requires knowledge of the resistance and the accurate adjustment of the variable resistor.

An alternative system is to use chosen resistors for R_1, R_2 and R_3. Instead of balancing the circuit, we can calculate the resistance of the component (in place of R_4) by mathematical analysis of the potential on the voltmeter.

This system can be used as a strain gauge, to produce a digital scale, or as a thermometer by using a thermistor.

Power

In order to make circuits work we need to supply them with enough energy. To determine this we need to know what energy they will require per second. This is what we call **power**. Power is a measure of the amount of energy required each second and is given a special unit – the **watt** (W). One watt is equal to 1 joule per second.

For Interest	James Watt

The unit of power, the watt, is named after James Watt. He was responsible for a radical improvement to the steam engine, leading to vastly improved efficiency, and he was highly influential in Scottish engineering. Both Heriot Watt University and the James Watt Building at the University of Glasgow bear his name.

Earlier, we learned that potential difference is measured in volts, where $1\,V = 1\,J/C$, and current is measured in amperes, where $1\,A = 1\,C/s$. If we multiply potential difference and current together we get $J/C \times C/s$. This

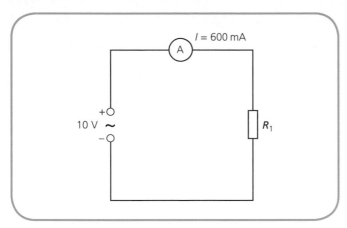

Figure 14.34

gives us JC/Cs, which, removing C, gives us J/s. This is 1 W. As a result, we can state that power (P) is

$$P = IV$$

An example of the use of this equation is shown below. Calculate the power in the resistor of the circuit shown in Figure 14.34.

$V = 10\,V$
$I = 600\,mA$
$P = ?$
$P = IV$
$P = 10 \times 600 \times 10^{-3} = 6\,W$

To calculate the power, we do not need to calculate the resistance of the circuit; instead we can concentrate on the current and the potential.

We also know that $V = IR$, and substituting this into the previous equation, we can say that

$$P = IV = I \times IR = I^2R$$

Similarly, $I = V/R$ so

$$P = IV = (V/R) \times V = V^2/R$$

This leaves us with three separate equations for power. These can be incredibly useful when we are trying to investigate various circuits with limited information.

$$P = IV = I^2R = V^2/R$$

Investigating circuits

We are now going to look at two different circuits which we need to investigate. You may be asked to determine:

- the resistance of any component
- the potential across a section of a circuit
- the current through any branch
- the power being used by a circuit or an individual component.

As with all questions of this type – it is important to break the question into small sections in order to tackle them individually.

Worked Example 14.7

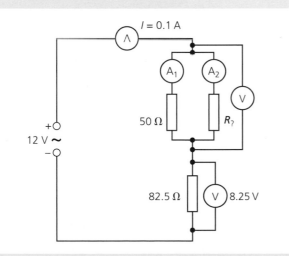

Figure 14.35

From the circuit in Figure 14.35 calculate

a) the potential across the parallel section

b) the readings on ammeters A₁ and A₂

c) the value of the unknown resistor

d) the total power used in this circuit.

a) A large amount of information is required from this circuit, so we start with the simplest part. As we have V_s and the potential across the lower resistor, we can deduce the potential across the parallel section as $V_p = 12 - 8.25\,V = 3.75\,V$.

b) We are now in a position to calculate the reading on A₁ because we know the resistance and potential for this branch.

$V = 3.75\,V$
$I = ?$
$R = 50\,\Omega$
$V = IR$
$3.75 = I \times 50$
$I = 3.75/50 = 0.075\,A$

So the reading on A₁ is 75 mA.

We know that the whole circuit has a current of 0.1 A from the supply current.

The total currents in parallel branches must add up to the total current which is supplied. So we can calculate the reading on $A_2 = 0.1 - 0.075 = 0.025\,A = 25\,mA$

We have calculated the potential across the parallel section and the currents through each branch. We still need to calculate the value of the resistor and the total power.

c) For the second branch:

$V = 3.75\,V$, because the potential across each branch of a parallel section is the same
$I = 0.025\,A$ as calculated in part **b)**
$R = ?$
$V = IR$, so $3.75 = 0.025 \times R$
$R = 3.75/0.025 = 150\,\Omega$

d) To find the power:

$P = ?$
$I = 0.1\,A$, as it is the total power of the whole circuit
$V = 12\,V$
$P = IV = 0.1 \times 12 = 1.2\,W$

Worked Example 14.8

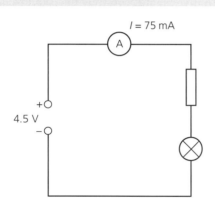

Figure 14.36

The lamp in this circuit requires 1.2 V across it in order for it to work at the desired level, but must not reach more than 4 V or it will be damaged. The resistance of the lamp is 10 Ω.

a) Calculate the value of the resistor.

b) Does the lamp have the required potential?

c) Redesign the circuit so that the lamp brightness can be altered easily. Suggest values for all components.

a) We know the supply voltage and the current, so we can calculate the total resistance of the circuit.

$V = 4.5\,V$
$I = 75\,mA$
$R = ?$
$V = IR$
$4.5 = 0.075 \times R$
$R = 4.5/0.075 = 60\,\Omega$

We know that the lamp has a resistance of 10 Ω and that the resistances in a series circuit add up to the total resistance, so $R_1 = R_T - R_{lamp} = 60 - 10 = 50\,\Omega$

b) Using Ohm's Law we can calculate the potential across the lamp.

$R_{lamp} = 10\,\Omega$
$V = ?$
$I = 75\,mA$
$V = IR$
$V = 0.075 \times 10 = 0.75\,V$

The potential across the lamp is too low.

c) If we want to be able to alter the potential across the lamp, we need to be able to vary the resistance of the whole circuit. The best way to do this is with a variable resistor. The question asks for values so we must consider the values we may wish to go between. We know that the lamp should not reach 4 V so the maximum should be 3.99 V. At the lower end, the lowest value should be 1.2 V. Using these values we can reach the maximum and minimum values for the variable resistor.

We can use the potential divider equation because we know R_2, V_2, V_S and this will enable us to calculate R_1.

$V_{2max} = 3.99\,V$
$V_{2min} = 1.2\,V$
$V_S = 4.5\,V$
$R_2 = 10\,\Omega$

For the maximum V_2:

$V_{2max} = \dfrac{R_2}{(R_1 + R_2)} \times V_S$

$3.99 = \dfrac{10}{(R_1 + 10)} \times 4.5$

$3.99(R_1 + 10) = 45$

$(R_1 + 10) = \dfrac{45}{3.99} = 11.28$

$R_1 = 11.28 - 10 = 1.28\,\Omega$

For the minimum V_2:

$V_{2min} = \dfrac{R_2}{(R_1 + R_2)} \times V_S$

$1.2 = \dfrac{10}{(R_1 + 10)} \times 4.5$

$1.2(R_1 + 10) = 45$

$(R_1 + 10) = \dfrac{45}{1.2} = 37.50$

$R_1 = 37.50 - 10 = 27.50\,\Omega$

We can then redraw the circuit with the new component in place.

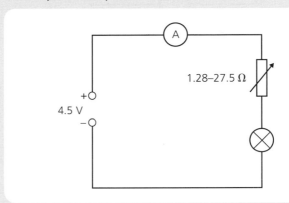

Figure 14.37

Questions

8 What is the definition of the volt?

9 A $10\,\Omega$ resistor is connected to a $50\,V$ supply. What current is passing through the resistor?

10 Two students are investigating the following circuit.

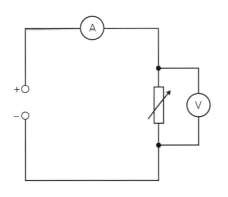

Figure 14.38

They are able to control the resistance and the potential difference of the circuit. Copy and complete Table 14.1.

Potential difference/V	Resistance/Ω	Current/A
5	2.5 k	
4		5×10^{-3}
	27.5	0.8
230	23	
	2	0.5
0.5		0.1×10^{-3}
	80	1.25

Table 14.1

11 In each of the following examples, calculate the total series resistance.

a)

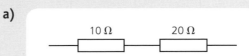

Figure 14.39

b)

Figure 14.40

c)

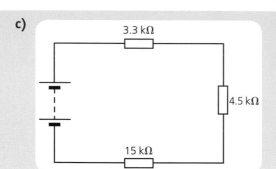

Figure 14.41

d)

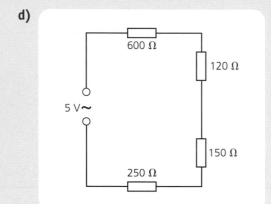

Figure 14.42

12 In each of the following examples, calculate the total parallel resistance.

a)

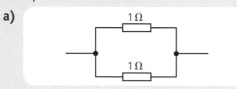

Figure 14.43

b)

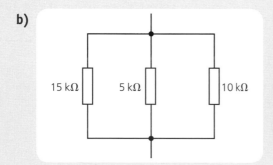

Figure 14.44

c)

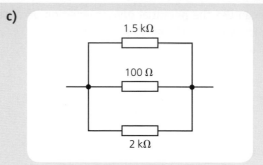

Figure 14.45

d)

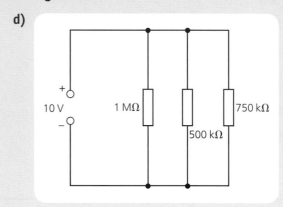

Figure 14.46

13 In each of the following examples, calculate the total resistance.

a)

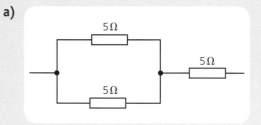

Figure 14.47

b)

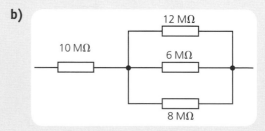

Figure 14.48

c)

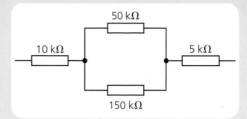

Figure 14.49

d)

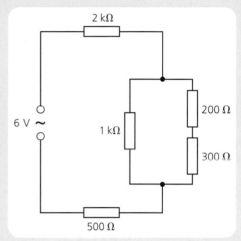

Figure 14.50

14 In each of the following examples, fill in the missing resistor value.

a)

Figure 14.51

b)

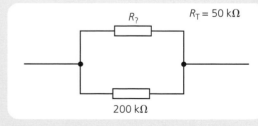

Figure 14.52

c)

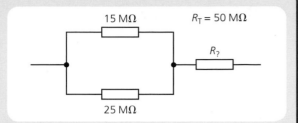

Figure 14.53

d)

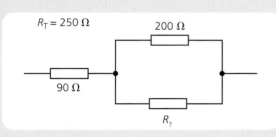

Figure 14.54

15 Calculate the supply p.d. in each of these circuits.

a)

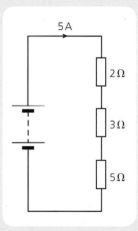

Figure 14.55

b)

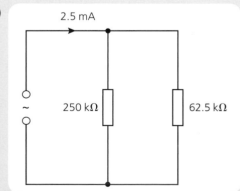

Figure 14.56

c)

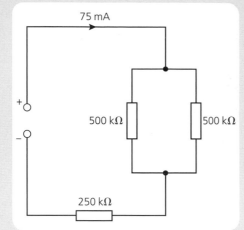

Figure 14.57

d)

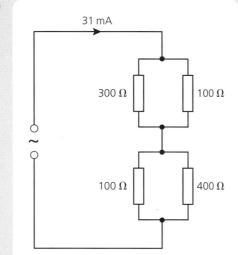

Figure 14.58

16 Calculate the readings on the ammeters in these circuits.

a)

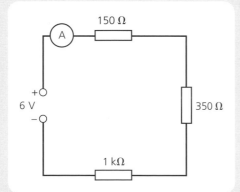

Figure 14.59

b)

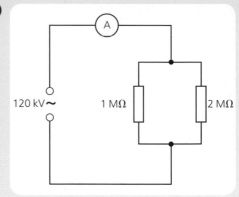

Figure 14.60

c)

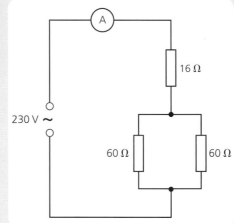

Figure 14.61

d)

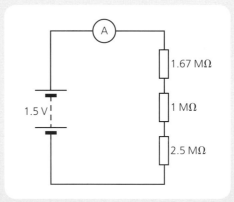

Figure 14.62

17 Calculate the readings on the voltmeters in each of these circuits.

a)

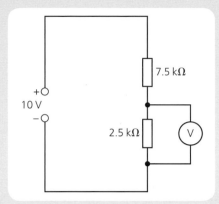

Figure 14.63

b)

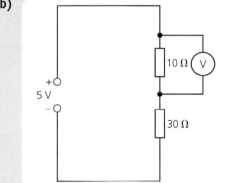

Figure 14.64

c)

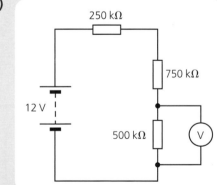

Figure 14.65

d)

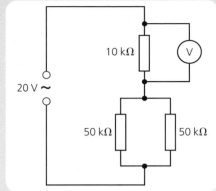

Figure 14.66

18 Calculate the value of the supply voltage in each of the following circuits.

a)

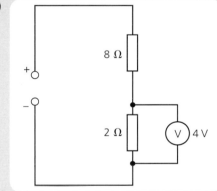

Figure 14.67

b)

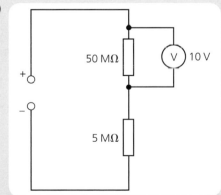

Figure 14.68

c)

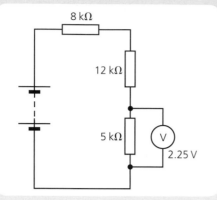

Figure 14.69

d)

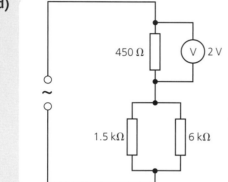

Figure 14.70

19 What value of resistor is required for a 2V lamp to light in a circuit which includes a 5V supply, a resistor and a lamp of resistance 4Ω?

20 You have four 1kΩ resistors and a 10V supply. Design a circuit which has a 4V p.d. across one of the resistors and place a voltmeter in the correct position.

21 The lamp in each of these circuits is rated as 8Ω. Calculate the power used in the lamp in each case.

a)

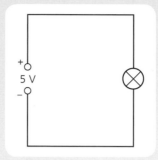

Figure 14.71

b)

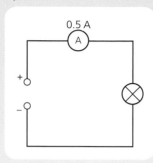

Figure 14.72

c)

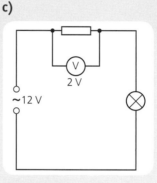

Figure 14.73

d)

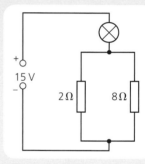

Figure 14.74

22 A student is tasked with making a circuit with four lamps as bright as possible with a 6V supply. Two of the lamps are 4Ω and the other two are 12Ω.

a) What is the best design for this circuit? Explain your answer.

b) How much power would be dissipated by the circuit?

Electrical sources and internal resistance

Internal resistance and EMF

Internal resistance

Resistance was discussed earlier and it was stated that every component in a circuit will have resistance. In reality this includes the power supply and this is true whether it is a mains power pack or a battery. Up to this point we have ignored this when carrying out calculations, but in some cases it is incredibly important as it will change the operation of some low-resistance circuits.

We refer to the resistance of the power supplies as **internal resistance** and assign it the symbol r. Internal resistance is measured in ohms. In most modern batteries it is usually about one or two ohms but we still need to analyse its impact because if the load (any components) in our circuit is of a similar size, the internal resistance will have a major impact on the current in the circuit.

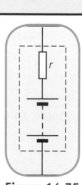

Figure 14.75
A battery has an internal resistance. We give internal resistance the symbol r, and measure it in ohms

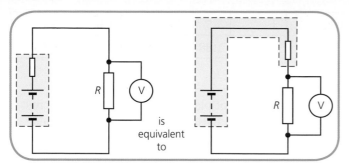

Figure 14.76 These two circuits are the same, you will notice that the second one looks just like a normal potential divider from earlier

EMF

The power supplies we use – mains and batteries – all supply energy to the circuit by changing some other form of energy into electrical energy. We define the total electrical energy supplied to the charges in the circuit as the **electromotive force** which we will refer to as EMF. This is measured in volts. It is essential at this point to note that while we call it the electromotive 'force', EMF is not a force – the name is a result of historical understanding. We will refer to it from here on as EMF and denote it by the letter E.

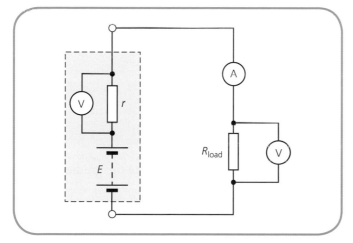

Figure 14.77 The EMF (or E) is made up of the potential difference across r and the potential difference across the load resistor

Figure 14.77 shows a simple series circuit which has a battery in series with a single load resistor. The battery has an internal resistance. We know that the current would be the same at any point in the circuit – even if we could measure it between the battery and the internal resistance – so we can state the following:

The potential across the internal resistance can be calculated using Ohm's Law and is $I \times r$.

The potential across the load resistance is $I \times R_{load}$.

The EMF across the battery is the total of the potential across each of the resistors. In equation form this is written as:

$$E = IR_{load} + Ir \quad \text{or} \quad E = I(R_{load} + r)$$

The potential 'used' across the internal resistance is not able to be used in our circuit. As such we could refer to it as 'lost' potential. We give it the label V_{lost}.

The potential across the load resistor is the same potential that we would measure across the battery. We can measure this at the terminals of the battery. For this reason we give it the name terminal potential difference, or V_{tpd}.

So, $E = V_{tpd} + V_{lost}$

Or, as it is often written

$$E = V_{tpd} + Ir$$

Required experiment: Investigating EMF and r

In order to determine EMF and r we need to design a circuit where we can vary the load resistor.

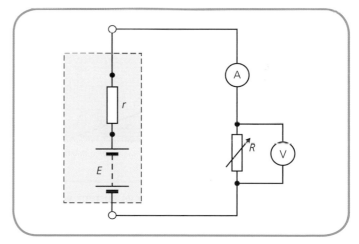

Figure 14.78 Circuit setup which allows the measurement of EMF and internal resistance

If we take multiple measurements using this circuit, we can determine both the EMF and the internal resistance. The apparatus is set up and the variable resistor is set to its maximum value. The readings are noted and the resistance altered steadily to a range of values. For each value of current the voltmeter reading is taken. Typical results are shown in Table 14.2.

Current/A	1	1.5	2	2.5	3	3.5	4	4.5	5
Terminal potential/V	3.74	3.39	3.00	2.62	2.24	1.88	1.51	1.03	0.48

Table 14.2

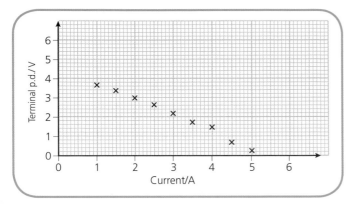

Figure 14.79 Graph of terminal p.d. vs current

It is difficult to spot a trend or pattern with data in this form, so a graph may be of use (Figure 14.79).

The graph shows that as the current increases, the terminal p.d. decreases. The next stage is to add a line of best fit to show this trend, extending it on both sides.

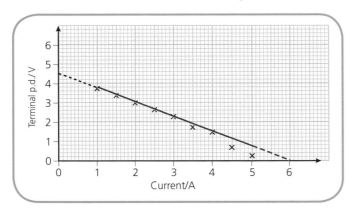

Figure 14.80 Graph of terminal p.d. vs current with a line of best fit added

The extensions to the graph allow us to understand important features of the circuit. When there is zero current flowing in the circuit, the graph (Figure 14.80) shows that V_{tpd} would be 4.5 V. This is the reading when no current is being drawn from the battery. This is the EMF of the supply. If no current flows, the effect of the internal resistance is nil.

At the other end of the line we have a current of 6 A but there is no terminal p.d. This means that the internal resistance has 'used' all the potential from the supply. This is the maximum current the battery can produce.

It is the equivalent of placing a conducting wire across the terminals of the battery (in other words, a short circuit). This is not a good thing to do as the battery dissipates all its energy very quickly and it may even damage it.

When there is a short circuit, the whole EMF is across the internal resistance. We can calculate the internal resistance using the EMF and the short circuit current, which we obtained from the graph.

In this case $V = IR$ can be replaced with $E = I_{sc}r$

So, $r = E/I_{sc} = 4.5/6 = 0.75\,\Omega$

There is an alternative graphical method for calculating the internal resistance using $E = V_{tpd} + Ir$. We can rearrange this to give $V_{tpd} = E - Ir$. Rearranging a bit more gives us $V_{tpd} = -rI + E$. This is similar to $y = mx + c$.

If we plot a graph of V_{tpd} against I (Figure 14.81), m is equivalent to $-r$ and E is equivalent to c.

From the graph,

$V_{tpd} = 3\,V$ when $I = 2\,A$

and

$V_{tpd} = E = 4.5\,V$ when $I = 0$

We can use these figures to calculate the gradient of the line: $(4.5 - 3)/(0 - 2) = -0.75$.

But gradient of the graph is $-r$. Hence $r = 0.75\,\Omega$.

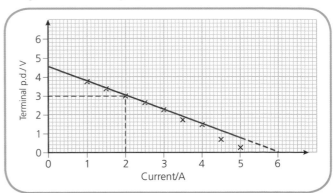

Figure 14.81 The values in the graph of terminal p.d. vs current can be used to calculate the EMF of the supply

The intercept on the y-axis is 4.5 V. Hence $E = 4.5\,V$.

$$E = V_{tpd} + Ir$$

So $E - V_{tpd} = Ir$

$$r = \frac{(E - V_{tpd})}{I} = \frac{(4.5 - 3)}{2} = 0.75\,\Omega$$

Load matching

In order to get the maximum power output from a circuit we need to match the load and internal resistance. To investigate this, we will have a look at a simple circuit (Figure 14.82).

If we want to get the maximum power possible through the lamp, we need to get the correct resistance.

The results in Table 14.3 clearly show that when $R_{lamp} = r$, the power in the lamp is the largest it can be. This is not to be mistaken for the most efficient circuit. The internal resistance is incredibly small and this means that if we want to match it, the current must be very high – as can be seen from the table. The resistance of a lamp is not going to be as small as just $1\,\Omega$. In real circuits, however, there is a way round this problem. While we cannot increase the resistance of the battery, we can put another resistor in series with it.

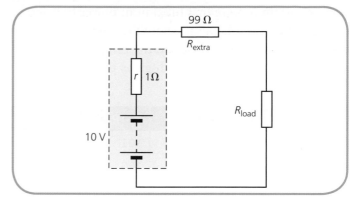

Figure 14.83

Since $I = V/R$, when we increase resistance, we decrease current and this is often desirable since current can make devices incredibly hot. For example, 5 A in a modern computer processor would burn it out completely.

In Figure 14.83 the load resistance – which could be a lamp or a motor or any other component – will have the maximum power when $R_{load} = r + R_{extra}$. If we repeat the experiment from before, we can see this in action.

Table 14.4 now shows that the maximum power output is when $R_{load} = r + R_{extra}$. The current and power dissipated have also decreased by more than an order of magnitude. Circuit designers have two reasons for placing the extra resistor in their circuits:

1 It means that higher resistance components can be 'load matched'.

2 If the extra resistance is significantly higher than the internal resistance, the circuit operation is no longer dependent on the internal resistor since matching the extra resistance is more important.

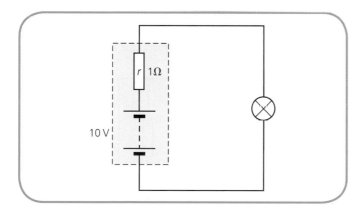

Figure 14.82

r/Ω	R_{lamp}/Ω	$(R_{lamp} + r)/\Omega$	V_{lamp}/V	I/A	P_{lamp}/W
1.00	0.50	1.50	3.34	6.67	22.28
1.00	0.75	1.75	4.28	5.71	24.45
1.00	1.00	2.00	5.00	5.00	25.00
1.00	1.25	2.25	5.55	4.44	24.64
1.00	1.50	2.50	6.00	4.00	24.00

Table 14.3

It is important to note that maximum power transfer does not occur at the point where the circuit is most efficient. For the best efficiency, the ratio of load to source resistance should be as high as possible.

$r + R_{extra}/\Omega$	R_{load}/Ω	R_T ($R_{load} + R_{extra} + r$)/Ω	V_{load}/V	I/mA	P_{load}/mW
100	1	101	0.10	99	9
100	50	150	3.33	67	223
100	100	200	5.00	50	250
100	150	250	6.00	40	240
100	200	300	6.67	33	220

Table 14.4

For Interest   Open circuits and short circuits

Open circuits and short circuits are important definitions when we are discussing circuitry. They may be intentional or due to an error in a circuit.

An open circuit occurs when there is a gap in a circuit. For example, when a switch is 'open' or 'off', this is essentially an open circuit. An open circuit can be thought of as having an infinite resistance. The example in Figure 14.84 shows an open circuit between terminals A and B. In this case, these wires may have been left there intentionally for an extra resistor to be added later.

A short circuit occurs when a wire connects both sides of the power supply. This can occur because a wire has been placed across the terminals but also if components fail or have been incorrectly attached to the circuit. It is very rare that this would be an intentional feature of a circuit as it is likely to cause an incredibly large current flow and there is a significant risk of overheating or even fire. Figure 14.85 shows an example of a short circuit.

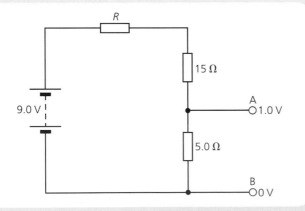

Figure 14.84 There is an open circuit between terminals A and B

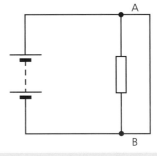

Figure 14.85 There is a short circuit between terminals A and B. The battery is connected to itself by this wire, meaning that the maximum current available will flow and the wire will get very hot

For Interest   The maximum power theorem (for those who like the proof)

It may come as a surprise that the maximum power occurs when we use exactly the same resistance for our load as the internal resistance of the circuit.

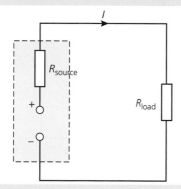

Figure 14.86

In the circuit shown in Figure 14.86, R_{source} is any internal resistance, plus any additional resistance placed in the circuit when it was designed. R_{load} is the load of the resistance – this could be any component.

The proof of the theorem is as follows:

$E = IR$

where E is the EMF of the supply and R is the total resistance including the internal resistance.

$$I = \frac{E}{R} = \frac{E}{(R_{source} + R_{load})}$$

$P = I^2R$

$P_{load} = I^2 R_{load}$

$$= \left(\frac{E}{R_{source} + R_{load}}\right)^2 R_{load}$$

$$= \left(\frac{E}{R_{source} + R_{load}}\right)\left(\frac{E}{R_{source} + R_{load}}\right) R_{load}$$

$$= \frac{E^2}{\frac{R_{source}^2}{R_{load}} + 2R_{source} + R_{load}}$$

The maximum P_{load} will be when the bottom line of the fraction is at a minimum. To get this, we must differentiate it.

$$\frac{d}{dR_{load}}\left(\frac{R_{source}^2}{R_{load}} + 2R_{source} + R_{load}\right) = 1 - \frac{R_{source}^2}{R_{load}^2}$$

The derivative of a function is 0 at a minimum, so

$$1 - \frac{R_{source}^2}{R_{load}^2} = 0 \qquad so \quad 1 = \frac{R_{source}^2}{R_{load}^2}$$

$$R_{load}^2 = R_{source}^2$$

$$R_{load} = R_{source} \text{ (assuming both are positive)}$$

This is a difficult proof – but if you follow it through you can see that when the resistances match, $P = V^2/R$ must be at its maximum, where V is the potential difference across the load.

Questions

Current/A	0.5	1	1.5	2	2.5	3	3.5	4	4.5
Terminal potential/V	4.21	3.65	3.22	2.71	2.13	1.74	1.20	0.75	0.18

Table 14.5

23 A student performs an investigation of internal resistance in a circuit. Using the information given in Table 14.5, draw a graph and state the values of r and E for the circuit.

24 Calculate the value of r in the following circuits.

a)

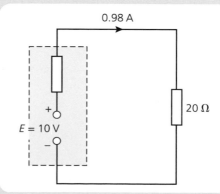

Figure 14.87

b)

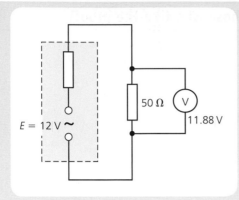

Figure 14.88

c)

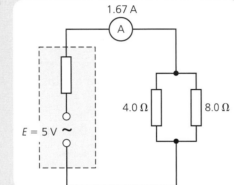

Figure 14.89

d)

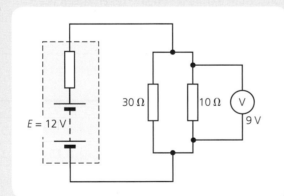

Figure 14.90

25 Calculate the value of the electromotive force in the following circuits.

a)

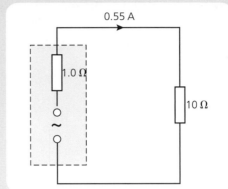

Figure 14.91

b)

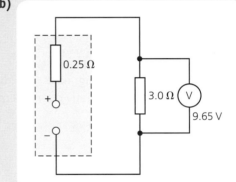

Figure 14.92

c)

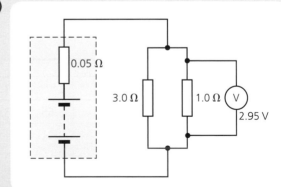

Figure 14.93

d)

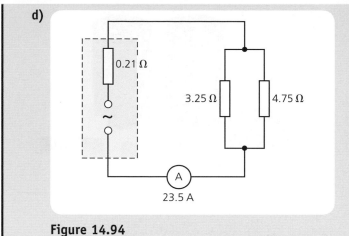

Figure 14.94

26 A manufacturer has been having difficulty with their power supply units (PSUs). An engineer recognises that the internal resistance has been inconsistent. The 30 V PSUs will be used to power 40 Ω motors which will not operate correctly at less than 24.5 V. What is the maximum internal resistance allowed for the PSU?

Capacitors

In the 1700s the phenomenon of electricity was the subject of a large amount of research and various experiments formed parts of spectacular shows, but very few practical uses had been found. Part of the problem was that of storage; there was no reliable means of storing electrical energy for any time. It had to be used as it was produced.

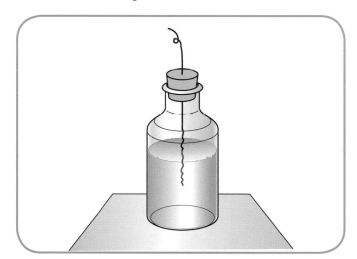

Figure 14.95 A simple Leyden jar

At this time there was a widely held belief that electricity was a fluid and so research focused on methods for capturing this fluid. The most obvious way was using a bottle as shown in Figure 14.95. The wire was connected to the static electricity generator and the electricity was thought to 'flow' into the glass bottle. Two scientists – von Kleist and van Musschenbroek – both worked on versions of this jar.

Dates suggest that von Kleist created the jar first. His experiment included holding the glass section of the jar while the wire was connected to the electricity supply. He then touched the wire and received a large electric shock. The experiment was repeated, but with the jar on an insulator. The results were very different. The water did not appear to be storing any of the electrical fluid any more. He was puzzled that he could not repeat his experiment. At around the same time van Musschenbroek realised that it was necessary to hold the jar – but could not yet explain why.

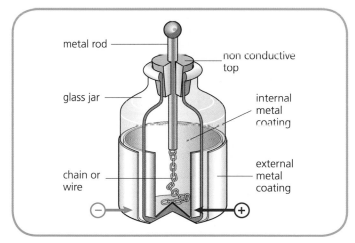

Figure 14.96 The addition of metal meant that the jar worked without the need to hold it, and could be connected to other electrical devices

There was continuous research into the phenomenon and eventually water was replaced by a metal coating on the inside of the jar. The outside was also covered in metal, meaning that the jar (now known as a Leyden jar) would work even if no one was holding it. This does not yet, however, tell us what caused the electricity to be stored and why the experimenters received a shock.

It can be explained if we use the modern concept of charge – since there is no such thing as electrical fluid. As the metal inside the jar becomes charged, it attracts an equal and opposite charge to the metal coating on the outside. In the earlier jars the experimenter's hand became oppositely charged and this meant that when they touched the wire, the charge moved quickly back to a neutral state or equilibrium – through the experimenter's body.

The glass in the jar is very important to this phenomenon as it stops the charge reaching the inner coating of metal but it is essential to realise that the charge is 'stored' in the metal and not in the glass. Leyden jars were the basis for the **capacitors** which are widely used today and allowed experimenters to store electrical energy for use in various experiments. They even created basic batteries by linking multiple Leyden jars.

It was later discovered that the glass was not required. The effect will happen with any two metal plates which are separated by poorly conducting materials – even air.

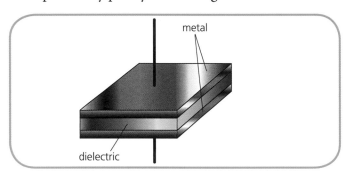

Figure 14.97 The basic form of a capacitor

A capacitor is essentially two conductors, usually metal, separated by a **dielectric** (a material which is usually an insulator – and therefore does not conduct well – but can sustain a large electric field across it) as shown in Figure 14.97. Widely used dielectrics include glass, ceramic materials, paper and plastics. Different dielectrics allow different amounts of energy to be stored in the capacitor.

Some practical examples of capacitors are shown in Figure 14.98. Capacitors are vital components in many circuits which we use every day, including computers. Variable capacitors can be made and these can be used in 'tuning' circuits for radio or as variable timers. All of them work on the basis that the two plates will become equally and oppositely charged.

Figure 14.98 Various capacitors

Capacitance

Every capacitor has a **capacitance**, in other words, the maximum amount of charge it can hold for a given potential difference. The capacitance depends on the materials which are used to make the capacitor but also on the area of the metal plates (larger plates mean more capacitance) and the thickness of the dielectric. For this reason, to create physically small capacitors, with large capacitance, capacitors are often made by turning the 'sandwich' of plates and dielectrics into a 'swiss roll' with an insulating layer to stop the plates coming into contact with each other, as shown in Figure 14.99.

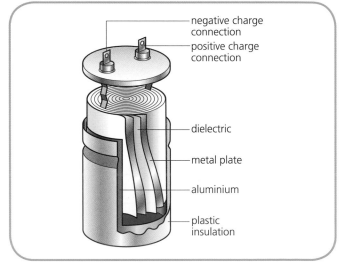

Figure 14.99 The plates and dielectric of a capacitor can be rolled up to make the capacitor much smaller

Now that we know what a capacitor is we can start to investigate its properties. So far we know that the plates of a capacitor will charge up with equal and opposite charge, but for this to happen there must be electrical energy in the circuit. The power supply of the circuit provides the energy and this is measured in volts.

The amount of electrical energy – and so the charge (measured in coulombs) – which can be stored in a capacitor directly depends on the potential difference across it. This is shown in Table 14.6 for a particular capacitor.

Supply potential difference/V	Maximum charge in the capacitor/μC
1.00	5.00
2.00	10.00
3.00	15.00
4.00	20.00
5.00	25.00

Table 14.6

We can plot these results on a graph to show the relationship (Figure 14.100) and there is a clear straight line with gradient equal to 5×10^{-6}. This is a measure of how many coulombs of charge we are able to create with one volt of potential across the capacitor; in other words QV^{-1}. We give this value a unit of its own – farads or F.

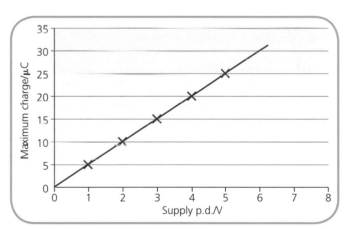

Figure 14.100 Graph of maximum charge vs p.d

Farads are the measure of capacitance: 1 F is 1 coulomb per volt. This would involve an incredibly large amount of charge and very rarely do we use capacitors of 1 F. In normal circuits, such as computers and televisions, we are far more likely to see capacitors ranging from 1 pF to 10 μF. The gradient of the line in our experiment was 5×10^{-6} and this is the capacitance of the capacitor used, 5 μF.

This leads us to the relationship between charge, capacitance and potential difference. The charge which

can be stored is the capacitance multiplied by the potential difference.

$$Q = CV$$

or rearranging this

$$C = Q/V$$

For Interest   **Michael Faraday**

The farad is named after Michael Faraday, a famous English physicist. He was famous for multiple discoveries through extensive experimentation, including some important work on electromagnetism.

It is important to note, however, that this relationship is not limitless. If the potential difference across the capacitor is too great, the electric field across the dielectric will 'break down'. In some cases, such as electrolytic capacitors, there could even be a small explosion!

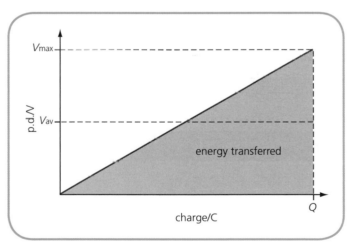

Figure 14.101 The energy stored in a capacitor can be calculated from the area below the line

As we discovered in Chapter 8, the work done, or energy transferred, is equal to the charge times the average potential difference, or $E_w = QV$ (where V is the average potential difference). This remains true in a capacitor. For this reason, we can look at the area under a Q vs V graph and calculate the total energy transferred. This graph will always look the same; the line will always form a triangle with the axes. To find the area under the graph, we use 'half base times height'. This would not be the average potential difference, though, it would be the maximum potential difference. So from the triangle area formula

$$E_w = \frac{1}{2}QV_{max}$$

And this is written simply as

$$E = \frac{1}{2}QV$$

We know another formula for Q ($Q = CV$) and we can substitute this into the energy formula to give us:

$$E = \frac{1}{2}CV^2$$

where C is the capacitance and V is the maximum p.d. across the capacitor in use.

The three forms of this equation, then, are

$$E = \frac{1}{2}QV = \frac{1}{2}CV^2 = \frac{1}{2}Q^2/C$$

We need to make sure that we keep the terms capacitance and charge clear in our minds, as the units can make it difficult.

- Capacitance is given by the letter C and is measured in farads, F.

- Charge is given by the letter Q and is measured in coulombs, C.

So if an equation has the letter C, it refers to capacitance. If, however, there is a measurement of 1 C, it is a measure of charge.

Worked Example 14.9

A circuit is created with a 20 nF capacitor and 1 kΩ resistor in series with a 12 V supply.

a) What is the maximum charge which can be stored in the capacitor?

b) How much energy is the capacitor storing when Q is at its highest?

   a) $Q = CV = 20 \times 10^{-9} \times 12 = 240\,nC$

   b) $E = \frac{1}{2}CV^2 = 0.5 \times 20 \times 10^{-9} \times 12^2$

      $= 10 \times 10^{-9} \times 144 = 1.44\,\mu J = 1.4\,\mu J$

Required experiment: Charging and discharging capacitors

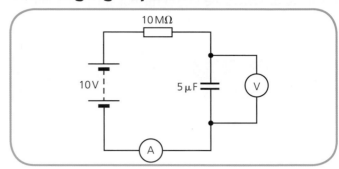

Figure 14.102 A simple RC circuit. The voltmeter reading will reach the supply potential, but may take some time to do so

Consider how a capacitor charges. The circuit in Figure 14.102 is set up and readings are taken every 50 s from the voltmeter. The results in Table 14.7 are obtained.

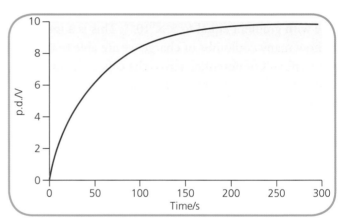

Figure 14.103 The potential difference across a capacitor as it charges

If we plot a graph of the results (see Figure 14.103), we can see that the capacitor charges more quickly at the start of the experiment. It can be seen that the potential across the capacitor has not reached 10 V. This is because the rate of charge transfer to the capacitor reduces as the capacitor stores more charge.

As the potential across the capacitor increases in the circuit, the potential across the resistor decreases. We know $V = IR$ and that current in a series circuit is the same at all points. This means that the current decreases as the voltage across the resistor decreases.

Time/s	0	50	100	150	200	250	300
p.d./V	0	6.32	8.65	9.50	9.82	9.93	9.98

Table 14.7

Time/s	0	50	100	150	200	250	300
Current/nA	1000	368	135	50	18	7	2

Table 14.8

If we look at the ammeter readings at the same time as the voltmeter readings above, we can produce Table 14.8.

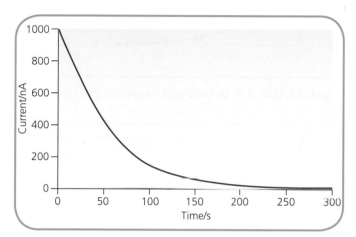

Figure 14.104 The current flowing in the circuit drops as the capacitor charges

The graph of these results is shown in Figure 14.104. The curve is the 'opposite' of the p.d. curve because the current falls as the p.d. rises. The reduction in current means that the capacitor charges more slowly.

The shape of these graphs (Figures 14.105 and 14.106) is always the same no matter what values we use for R, C and V, but the amount of time taken does change. As we have already discovered, most circuits will have more than one component in them and this means that they will always have a resistance. The resistance in a circuit inhibits the movement of charge and so reduces the current; this means that the charge takes longer to accumulate on the plates of the capacitor. The size of the capacitor matters too – the larger the capacitor, the longer it will take to fully charge, since it is capable of storing more energy.

A similar effect happens when we discharge the capacitor. Suppose the circuit from before is connected via a switch to a bulb as in Figure 14.107 on the next page.

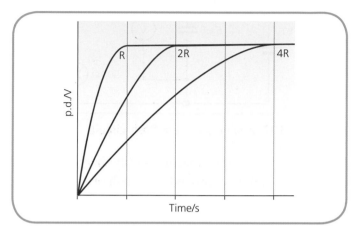

Figure 14.105 Increasing the resistance means that the capacitor takes longer to charge, but will not store any more energy

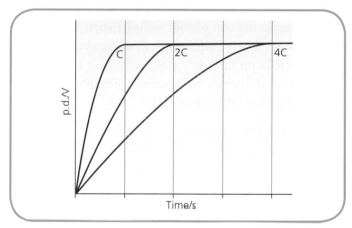

Figure 14.106 A larger capacitor takes longer to charge, but will store more energy

In this circuit the switch allows a connection between A, B or C and the capacitor. When A is connected, the capacitor will charge. When B is connected, the charge will be held in the capacitor, and when C is connected, the capacitor will discharge through the lamp. This is very similar to how the flash on a camera works.

Time/ms	0	2.5	5	7.5	10	12.5	15
p.d./V	10	3.68	1.35	0.50	0.18	0.07	0.02
Current/mA	20	7.36	1.70	1.00	0.36	0.14	0.04

Table 14.9

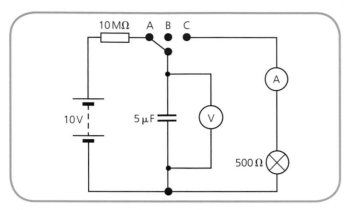

Figure 14.107 A simple camera flash circuit

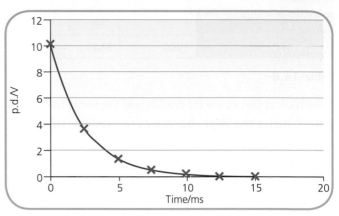

Figure 14.108 p.d. vs time as a capacitor discharges

Let us imagine that the capacitor is fully charged and that the switch is at B. The lamp is in the flash of a camera and we want to take a photograph. The switch is moved to C and the capacitor discharges as shown in Table 14.9.

Take careful note of the different scales in the discharge. The timescale is in ms and the current is in mA. Since the resistance of the lamp is much smaller than the resistor in the charging circuit, the capacitor discharges almost completely within 15 ms. This type of circuit allows a very bright flash as a large current flows through the lamp in a very short time. The graphs of the potential difference and current (Figures 14.108 and 14.109) show this more clearly.

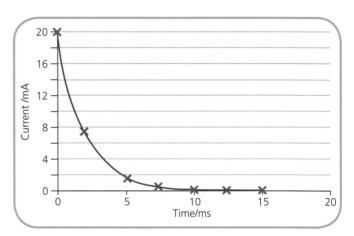

Figure 14.109 Current vs time as a capacitor discharges

For Interest   The time constant

The **time constant**, which we give the Greek letter τ (Tau), is a measure of how long a capacitor takes to charge and discharge. It is entirely dependent on the resistance and capacitance of the circuit.

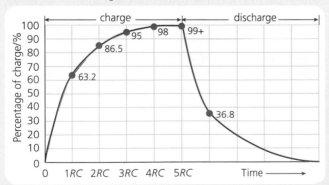

Figure 14.110 During the charging cycle, the potential across the capacitor increases by 63% of the potential across the resistor after 1τ. Similarly, it discharges by 63% of the potential after 1τ

If we multiply the capacitance and the resistance together we get a value which can be measured in seconds. So, for example, if there was a 1 mF capacitor and a 1 kΩ resistor, the time constant would be 1 s. This would be the time it took for the circuit to charge up to approximately 63% (1–1/e) of full potential. Similarly, it is how long it would take to discharge to approximately 37% (1/e) of the original potential. These numbers can be incredibly important when designing a circuit with an RC element, since this value will decide the circuit's operation. After 2τ, the capacitor will reach approximately 87% of maximum value of the potential, since it has now increased by 63% of the remaining charge. This is indicated on the RC curve in Figure 14.110 where a charge and discharge cycle is shown. In 1τ the capacitor charges up to 63% of the maximum p.d. Similarly, during discharge when $t = \tau$, the potential across the capacitor will have reduced to 37% of V at $t = 0$.

It is important to note that the shape of the current graph – large to small – is the same as for the charging cycle, but the p.d. curve has changed shape. If we increased the resistance of the lamp, the initial discharge would be small and would take longer. However with a lower resistance, the discharge would occur more quickly.

The capacitor in this circuit is acting as the power supply for the lamp. Unlike a battery, it is able to supply a large current very quickly and so it is ideal for an application which needs a short burst of current, but not as batteries for long-term storage.

Applications of capacitors

We have already seen that a circuit using the charging/discharging cycle of a capacitor can be used to supply the current for a camera flash. There are other uses too and it is not always desirable for them to discharge quickly.

Capacitors can be used to store energy for use if an alternative power supply shuts off. They will not be able to supply large currents for long, but in some applications such as the power supply to your computer, they can be used to stop the computer switching off in a 'brown out'. They can even be used to power small devices as batteries are being changed, to stop the loss of memory of, for example, your remote control.

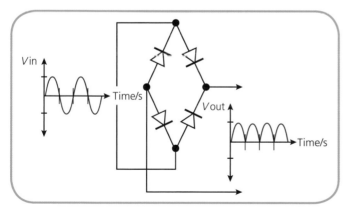

Figure 14.111 The AC signal feeds into the diode 'bridge' and as a result, the output is always positive

One major use of capacitors is in AC to DC converters. Diodes (which will be discussed in the next chapter) are used to convert the AC signal to a positive-only potential. This is shown in Figure 14.111.

The output of the circuit shown in Figure 14.111 is extremely uneven and would be no use in modern digital circuitry. The output of this circuit is fed into the circuit shown in Figure 14.112 and the capacitor has the effect, as it charges and discharges, of smoothing the signal, creating a much 'cleaner' signal which can be used as a DC p.d.

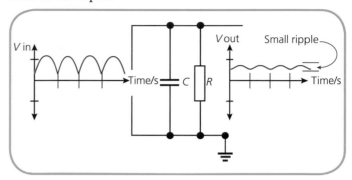

Figure 14.112 The additions of this RC circuit smooth the output of the diode bridge, creating a much flatter DC signal

Research Task

Find out how a capacitive touch screen works. These are widely used on phones and computers. How are they made? Why is a capacitive screen chosen? What are the alternatives?

For Interest   The theremin

Figure 14.113

A theremin is one of the earliest electrical musical instruments, developed by Leon Theremin in the early twentieth century. The player's hand is used as one plate of a capacitor. One hand controls the volume and the other controls the pitch. It is possible to build a theremin from some basic electronic components, but it is quite difficult to play!

Questions

27 A camera designer is making the circuit for a flash. She finds that charging is taking too long.

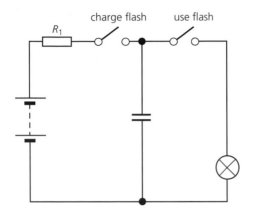

Figure 14.114

a) What could be changed in the circuit to make the capacitor charge more quickly?

b) The capacitor now charges quickly enough, but the flash needs to be a little brighter. What should be changed about the circuit in order to increase the flash, but maintain the time it takes to charge?

28 What is the maximum charge that can be stored in the capacitor in the following circuits?

a)

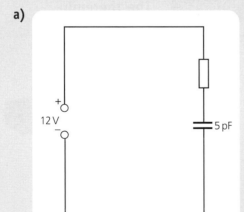

Figure 14.115

b)

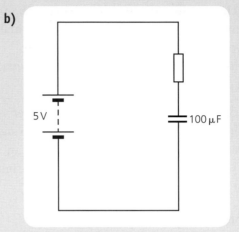

Figure 14.116

c)

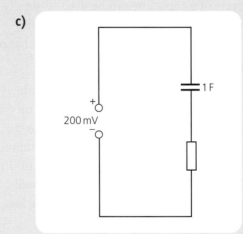

Figure 14.117

d)

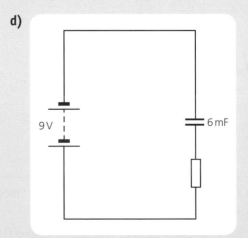

Figure 14.118

29 A student designs the following circuit. The results are not what she expected. Explain what has happened, why and what should be changed.

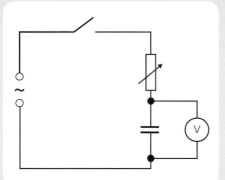

Figure 14.119

30 A circuit is made with a 12 V supply, a 1 Ω resistor and a 1 μF capacitor. The capacitor is charged and discharged. Sketch a graph of p.d. across the capacitor against time:

a) for the circuit described

b) if the resistor was changed to a 4 Ω resistor

c) if the resistor was 4 Ω but the capacitor was only 500 nF

d) if the circuit was returned to the original but the 12 V supply was doubled

e) if a variable resistor was used in order to keep the current constant in the original circuit.

31 Calculate the maximum energy which can be stored in the capacitors in these circuits.

a)

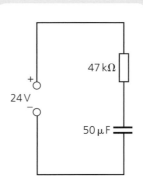

Figure 14.120

b)

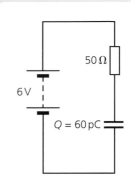

Figure 14.121

c)

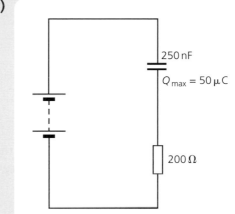

Figure 14.122

d)

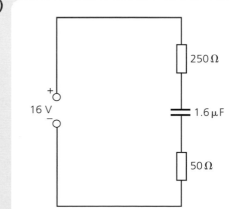

Figure 14.123

Consolidation Questions

1 The resistance between P and Q is 24 Ω.

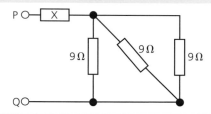

Figure 14.124

What is the resistance of X?

2 A 3.0 V lamp is connected to a 9.0 V DC supply as shown. The supply has negligible internal resistance.

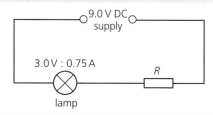

Figure 14.125

If the lamp is to operate at its correct rating, what must be the value of the resistor R?

3 In the potential divider circuit shown below, the variable resistor allows the potential difference between X and Y to be varied.

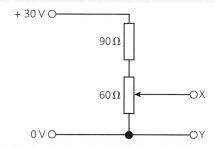

Figure 14.126

For the values given, what is the maximum potential difference which can be obtained across XY?

4 a) A circuit is used to provide a specific output potential difference, V_0, as shown in Figure 14.127. The circuit is supplied by a 12 V DC supply with negligible internal resistance.
i) Resistor X has a resistance of 2.2 kΩ. Vo is required to be 8.5 V. Determine the value of resistance which resistor Y should have.
ii) A third resistor is added to the circuit as shown in Figure 14.128. State what effect this

has on the potential difference across resistor Y. You must explain your answer.

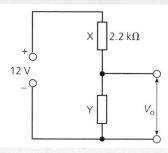

Figure 14.127

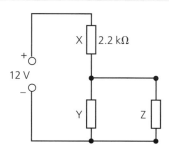

Figure 14.128

iii) The resistance of resistor Z is 3.3 kΩ. Determine the potential difference across resistor Z.

b) The circuit shown in Figure 14.129 is a type of bridge circuit. It can be used to compare the value of resistances. The bridge circuit is balanced when the reading on the voltmeter is 0. Determine the relationship between resistances A, B, C and D when the bridge is balanced.

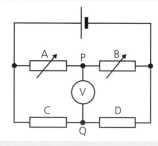

Figure 14.129

c) The resistances of fixed resistors C and D are each 60 Ω. Initially, the resistances of both variable resistors, A and B, are set to 60 Ω. The supply, which has negligible internal resistance, is a 10 V supply. The resistance of the variable resistors is now varied. Complete Table 14.10 to indicate the value on the voltmeter. You must include the sign.

Resistance of A/Ω	Resistance of B/Ω	Voltmeter reading/mV
60	60	i
61	60	−41
61	61	ii
61	62	iii
61	59	iv

Table 14.10

5 A particular alarm system operates when a switch is closed causing a lamp to light up. This causes a change in the potential difference across the connected battery, as shown in Figure 14.130. The battery has an EMF of 15.0 V and an internal resistance, r, of 0.30 Ω.

a) When the switch, S, is closed, the lamp has a current of 0.75 A through it. Determine the potential difference, V_1, when the switch is closed.

b) This circuit is now connected to a diode and capacitor as shown in Figure 14.131. State why, when switch S is closed, the potential difference across the capacitor does not decrease immediately. You must explain your answer.

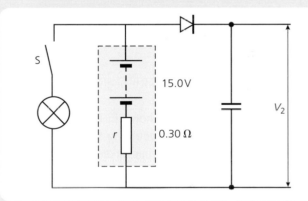

Figure 14.131

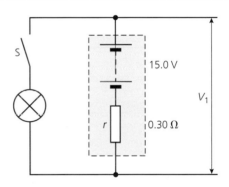

Figure 14.130

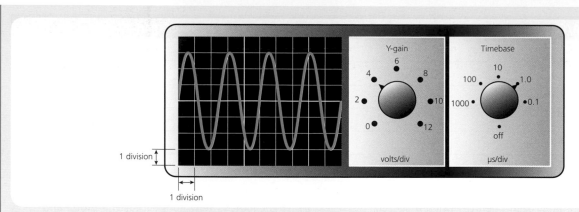

Figure 14.132

6 a) Figure 14.132 shows the trace on an oscilloscope which is connected to a signal generator.

Determine:
i) the frequency of the signal
ii) V_{peak}
iii) V_{rms}

b) A separate DC supply, with negligible internal resistance, is now connected to a resistor and capacitor in series as shown in Figure 14.133.
i) State the effect of doubling the resistance of the resistor on the time taken to charge the capacitor when the switch is closed.
ii) State the effect of halving the original resistance of the resistor on the energy stored in the capacitor when the switch is closed for a long time.
iii) State the effect on the maximum charge stored in the capacitor, and the time taken to reach maximum charge, if its capacitance is doubled but the original resistance of the resistor is halved.

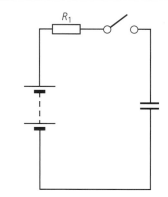

Figure 14.133

15 Electrons at work

Conductors, semiconductors and insulators

Conductors and insulators

So far, in physics, we have met two types of material: **conductors** and **insulators**. Conductors are materials which allow electrons to move, meaning that electrical current can 'flow' through them. In perfect insulators, the electrons are not free to move making it impossible for current to flow. To understand why, we need to look at individual atoms.

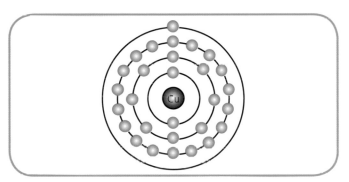

Figure 15.1 Structure of a copper atom. The electrons are contained at four different levels around the nucleus

Electrons in atoms are contained in energy levels. In Figure 15.1, we can see an example of this for copper, where there are four different levels. Each level can hold a certain number of electrons – the innermost can hold two electrons, the next eight and the third one can hold 18 electrons. There can be any number of electron levels, but no electron can be between levels. These levels can be empty, with no electrons, they may have some electrons, but not the maximum, or they can be full. If the level is full, then the electrons are very unlikely to move as these electrons are 'tightly' bound to the nucleus. In perfect insulators, the outermost level is full, so the electrons do not move. In conductors like copper, there are fewer electrons in the outermost energy level. These are not very tightly bound and will move away from the nucleus very easily. In order for a solid to conduct there must be 'free' electrons – those which are not tightly bound to the nucleus – and empty states in neighbouring atoms for the electrons to go.

Now we are familiar with the single atom, we need to look at what happens when a large number of atoms come together to produce a solid. Instead of looking at energy levels of each atom, we start to look at the **energy band** of the material. An energy band is a group of energy levels and, just like energy levels, the bands can be empty, full or partially full of electrons. No electrons can be in the space between the bands – the **band gap**. The size of the band gap is very important when we find out whether or not a material can conduct. A band is said to be occupied if there are electrons in it.

We will look at two main bands, the **valence band** and the **conduction band**. The highest-energy occupied band in an insulator is the valence band – it is full in a perfect insulator. This means that electrons cannot move into another position because there is no other position to move to. In a conductor, the band above this – the conduction band – has some electrons, though it is not fully occupied. In an insulator, there are no electrons in the conduction band.

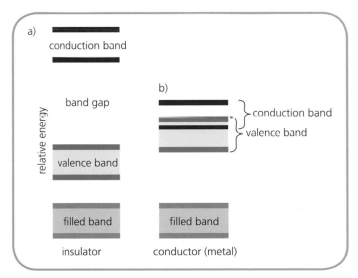

Figure 15.2 Energy bands in a) an insulator b) a conductor

Figure 15.2 shows the bands, and band gaps, of conductors and insulators at room temperature. The temperature is very important because we are dealing with energy – the more thermal energy in the material, the more energy the electrons have, making them more likely to move. At room temperature, there are some electrons with sufficient energy to exist in the

Research Task

There are various types of conductors and insulators. Different materials are used depending on what is required. Table 15.1 shows some examples and gives the **resistivity** of each material. (Resistivity is a measure of how resistive a material is, whereas resistance is a measure of how resistive a particular piece of the material is.)

Why do we use so many different materials? What are the benefits? How do engineers decide what material is required?

In the previous section we talked about resistance. Why is this important? Think about different situations – power lines, computers, cables for your headphones, scientific equipment, household appliances. Discuss what materials you would use, then find out what is actually used and why.

Material	Type	Resistivity/Ω m
silver	conductor	1.6×10^{-8}
gold	conductor	2.2×10^{-8}
copper	conductor	1.7×10^{-8}
iron	conductor	9.6×10^{-8}
aluminium	conductor	2.8×10^{-8}
graphite	conductor	$1.2 \times 10^{-6} - 9.8 \times 10^{-6}$
glass	insulator	$1.0 \times 10^{10} - 1.0 \times 10^{14}$
porcelain	insulator	1.0×10^{15}
quartz	insulator	1.0×10^{17}
high density polyethylene	insulator	1.0×10^{19}
polyvinyl chloride (PVC)	insulator	$1.0 \times 10^{12} - 1.0 \times 10^{16}$
fibreglass	insulator	8.6×10^{16}

Table 15.1

conduction band and so they can move throughout the material. Additionally, because the conduction and valence bands overlap, some electrons may acquire enough energy to be promoted to the conduction band and they too will be able to move throughout the material. This movement of electrons is the electrical current and so the material is known as a conductor. In insulators, however, the band gap is so large that at room temperature, it is very unlikely that an electron will have enough energy to move to the conduction band, so none of the electrons can allow current to flow. Insulators can be made to conduct, if enough extra energy is present. We will look at the effects of this in the next section.

Applications

One of the most obvious applications you will see of insulators and conductors together is in the power lines all around you. It is important that these are well insulated, because the energy involved is very high. It is also essential that they are efficient conductors, because they often transfer electrical energy over long distances. The cables need to be light because they are held up in the air between pylons. For this reason, copper is rarely used. Instead, the cables tend to be made of thin strands of aluminium since it is a good conductor, but also light, with some strands of steel for strength. The use of the thin strands, instead of thick single cables, also helps with something known as the **skin effect**.

For Interest | The skin effect

The skin effect is a phenomenon where alternating current will flow mainly on the outside of a conductor. The effect is greater at higher frequencies – that is one of the reasons mains voltage in Scotland is 50 Hz.

As the frequency gets higher, the layer where most of the current flows gets thinner and so the effective resistance of the material gets higher. Often cables will be made of lots of thinner strands instead of one large one. This is important when dealing with large currents, because increasing the diameter of a cable will not always decrease the resistance!

Figure 15.3

These cables have to be suspended and since the lines are at such high potential, it is essential that the insulators can withstand this. Suspension insulators are the most widely used insulator. Their job is to hold the conducting cable in the air, while also insulating the pylon. They are often made of glass or porcelain, but polymers are becoming more widely used.

Summary of conductors and insulators

- Every atom has electrons.

- Electrons occupy fixed energy levels.

- When atoms combine to create a solid, energy levels can be combined into bands.

- In insulators the valence band is full, but the conduction band is empty.

- The band gap of an insulator is large at room temperature.

- The conduction band must have some electrons in it to allow electrical conduction.

- In conductors, the valence band is full and there are some electrons in the conduction band.

- The band gap of a conductor is small at room temperature, so electrons can move between the bands.

- A conductor conducts well at room temperature.

Figure 15.4 Cables used in domestic properties are very well insulated

Figure 15.5 Ceramic discs are used on overhead pylons to provide the insulation

- At room temperature there is no electrical conduction in an insulator.

- Different materials will be used to conduct depending on various attributes including cost, resistivity and application.

Breakdown voltage and lightning

In the previous section, we investigated conductors and insulators. We learned that at room temperature, there is no conduction in a perfect insulator. Unfortunately, insulators are not always 'perfect'. In engineering, the manufacture of products always involves compromises. When making a product, availability, price and other characteristics such as flexibility or mass can be just as important as the insulating properties of an insulator. As a result the material will have properties which mean that there are limits to its ability to insulate.

The point at which an insulator starts to conduct is called the **breakdown voltage**. In order for an insulator to conduct, the electrical potential or voltage across it must be great. The increase in the energy of the electrons is so great that they are able to break free from their levels. Changes can occur in the material at this point which will not automatically reverse. Cracks in solid insulators may occur as the structure is weakened or heating may result in the material melting.

When deciding what insulator will be used for a product, the breakdown voltage will be noted. The problem is

Figure 15.6 High-voltage dielectric breakdown within a block of plexiglas

Figure 15.7 Lightning – a dramatic natural example of electrical breakdown

that it is very difficult to define; tiny differences in the material will change the breakdown voltage. Instead of trying to define an exact number where breakdown will occur, manufacturers will rate the maximum voltage where the insulator is guaranteed to work.

Everyone has seen the natural occurrence of electrical breakdown in a lightning storm. Air is not normally a conductor; however, when an electrostatic build up occurs in clouds, the potential difference can become so great as to cause lightning.

While the current in lightning is usually between 5 kA and 20 kA, currents of as much as 200 kA have been recorded. If lightning strikes silicon sand, the current creates such an enormous temperature that it can melt the sand instantly, turning it into fulgurites.

The sound of thunder which normally accompanies a lightning storm is caused by the heating effect. The movement of the electrons heats the air around the lightning stream. This occurs so quickly that the rapid expansion of the air creates the thunder. The feeling of fresh air after a thunderstorm is caused by the ionisation of the air which occurs. There are many types of lightning. Some bolts of lightning never reach the ground – they are simply transfers of electrons through a cloud. Sometimes the electron stream is travelling from the ground to the cloud – not the other way round. Even though it has been studied for a long time, the cause of lightning is still not fully understood.

For Interest   Fulgurites

Fulgurites are hollow tubes of glass caused by the rapid heating when lightning strikes silica sand. The tubes can be over 4m in length and come in various colours. The colour depends on the other materials surrounding the sand.

Figure 15.8 Fulgurites

Research Task

Harnessing the power of lightning

There is an incredible amount of energy in lightning. Investigate how much energy there is and find out about attempts which have been made to harness this power. How would you go about it? Think about a system which would allow safe collection of energy from lightning. Are there any problems you can think of? Would lightning be useful as the sole supply for a country?

Semiconductors

We now know much about conductors and insulators – but there is something in between. **Semiconductors** are a special set of materials which provide the intermediate step. Silicon is the most popular semiconductor – it is used for most computer processors and other chips which you will find in phones – but different materials can be used because they have properties which are similar and useful.

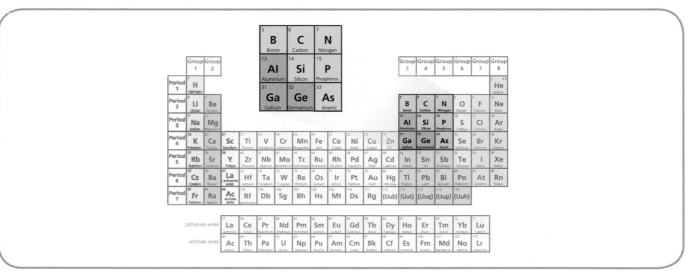

Figure 15.9 Semiconductors in the Periodic Table. Most devices which are in common use are based on group IV elements – in particular silicon and germanium – but mixing elements from groups III and IV, such as gallium and arsenic, gives 'III–V' semiconductors such as gallium arsenide

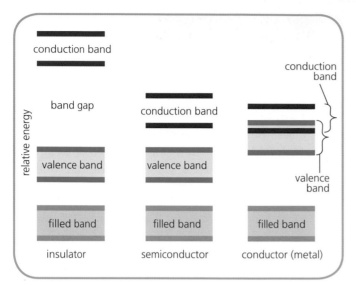

Figure 15.10 In a semiconductor, some electrons can occupy the conduction band at room temperature. This is because the band gap is more similar in size to that of conductors than insulators

At room temperature pure silicon has a resistivity of $1.0 \times 10^3 \, \Omega \, \text{m}$. This is 11 orders of magnitude larger than copper but is 12 orders of magnitude lower than porcelain. At this temperature current will flow, but only a small amount.

Application

The conductivity of a semiconductor is dependent upon temperature. The higher the temperature, the greater the conductivity. This can be utilised in an electrical component called a **thermistor**. As the resistivity of a semiconductor is somewhere between that of an insulator and a conductor, it can be used to make a special type of resistor. Instead of having a fixed resistance, the resistance will vary depending on the temperature. As the temperature rises the electrons will have more energy and will be more likely to reach the conduction band. This, in turn, makes the movement of electrons easier and the resistance lowers.

Figure 15.11 The circuit symbol for a thermistor

Activities

Design a circuit which will light a lamp if the temperature goes above a certain level. Draw the circuit diagram and, if the materials are available, build the circuit. If there is time, investigate alterations to the circuit. Could you use the thermistor as a temperature gauge? Could you make the lamp light when the temperature drops instead?

Hall effect sensors

Semiconductors can also be used as sensors. When there is a current in a semiconductor placed in a magnetic field a potential difference is created perpendicular to the current and the magnetic field. This potential difference, or voltage, can be measured and the measurement used in circuits to sense the magnetic field. One major advantage of this type of sensor is the removal of moving parts, such as in a 'reed switch'; another is the fact that there is no electrical contact being made – and therefore no risk of sparking – which allows the sensor to be sealed. Can you think of anywhere we might use this sort of sensor? Where would we want information, but want to avoid contact between the sensor and the environment?

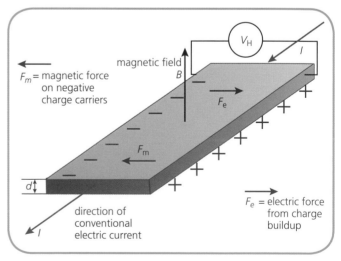

Figure 15.12 The Hall effect shows that if a material is carrying an electric current, a perpendicular voltage will be created across the material if a magnetic field is present. This is caused by the movement of charges to either end of the material. What is particularly amazing is that Edwin H. Hall discovered this effect in 1879, 20 years before the discovery of the electron!

p–n junctions

Doping semiconductors

The start of most semiconductor devices is with an ingot, as shown in Figure 15.13. These are 'grown' slowly using a seed crystal, similar to crystal-growing experiments you may have done in chemistry. These ingots are very long – up to 2 m – and are sliced into thin pieces of silicon (less than 1 mm thick) before being used to produce various electronic components.

Figure 15.13 This is a silicon ingot. Silicon is the most widely used semiconductor and is 'grown' in this way to produce a uniform crystal structure

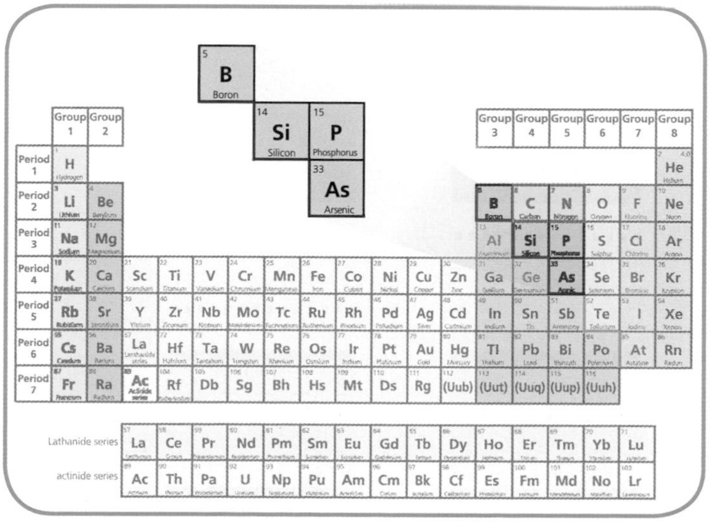

Basic silicon is not the end of the story though. To produce microchips for computers and phones, the silicon must be '**doped**'. Doping is a process whereby electrons are added or removed from the silicon structure, meaning that there are either fewer or more electrons than in standard silicon. This alters the conductivity of silicon, especially when voltage is applied to the devices.

There are two widespread methods of doping. The first is to introduce another element to the molten silicon before the ingot is grown. The second, and more popular, is to dope the silicon after the wafer has been cut with processes known as diffusion and ion implantation which are followed by an annealing (heating to a temperature which will make the material recrystallise once it begins to cool).

When we dope a semiconductor, we make it less pure. We call the additions 'impurities' and there are two types – acceptors and donors. To work out what these impurities will be, we need to look at the Periodic Table again (see Figure 15.14).

The most widely used semiconductor is silicon. It is a IV-type semiconductor, meaning that it has four electron bonds per atom. When a silicon crystal is 100% pure, the bonds between atoms are made up of two electrons – one from each atom – and these create a regular lattice, as shown in Figure 15.15. This form of silicon is called **intrinsic silicon**. We can alter this by adding the impurities to change the crystal structure.

For a IV-type semiconductor like silicon, we can add electrons using an element from the V group – usually

Figure 15.14 Currently, silicon is the most widely used semiconductor. It has four electron bonds per atom

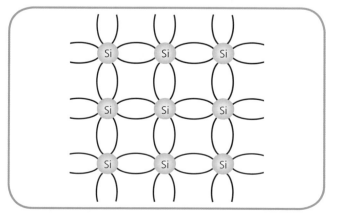

Figure 15.15 The crystalline structure of silicon, showing the double bonds between each atom in a perfect lattice

phosphorus or arsenic – which acts as a donor. To remove one of the electron bonds to leave a 'hole', we can use a group III element – usually boron – as an acceptor. These will result in crystal lattices as shown in Figure 15.16.

The two types of silicon which are created are **n-type** and **p-type**. The impurities added to the silicon mean that n-type has an extra electron for every donor and p-type gains a hole (a space where an electron

should be) for every acceptor. This has the effect of altering the conduction properties of the silicon but the silicon remains neutral overall.

The addition or removal of electrons has an effect on the positions of electrons in the semiconductors. In n-type the bands are lowered and electrons are more likely reach the conduction band. In p-type the bands are raised and holes are more likely

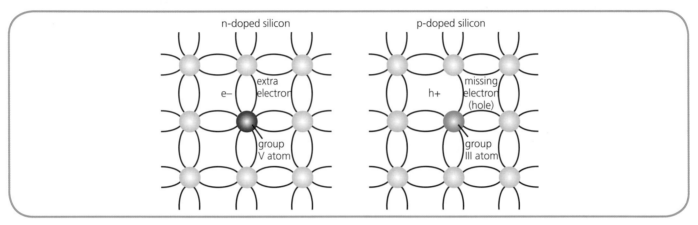

Figure 15.16 The crystalline structure of n- and p-type silicon

For Interest | Fermi level

Engineers and scientists use a special energy level to help them understand what is happening in semiconductors. The **Fermi level**, or E_f as you will see it written, simply denotes the point where it is equally probable that an electron is or is not present. The important factor about it is that the Fermi level must remain constant through any semiconductor

device if there is no applied potential. Figure 15.17 shows the energy band diagrams for intrinsic, n-type and p-type silicon. In n-type silicon, the Fermi level gets closer to the conduction band since electrons have been added. In p-type, as electrons have been removed, the Fermi level is closer to the valence band.

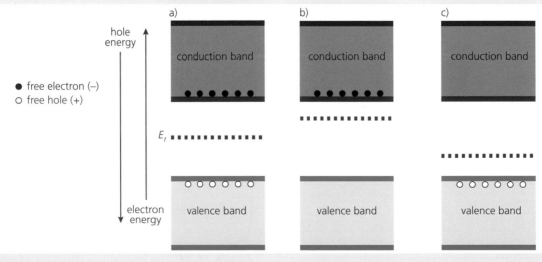

Figure 15.17 Energy diagrams for a) intrinsic, b) n-type and c) p-type silicon. The Fermi level is moved towards one of the energy bands as electrons are added or removed

For Interest   **Enrico Fermi**

The Fermi level is named after the Nobel Prize-winning scientist Enrico Fermi. He won the prize for discoveries in radioactivity and went on to be part of the Manhattan Project which was created to develop the atomic bomb.

to appear in the valence band. In both cases, the conductivity is increased.

p–n junctions

These doped semiconductors do not do very much on their own. To make useful devices, we must put different types together. One of the most widely used devices – present in many everyday electronic appliances – is the **diode**. A diode is a device which conducts electricity in one direction much more than in the other. For the moment, we will focus on the p–n diode (Figure 15.18).

When p-type silicon is interfaced with n-type silicon, as the bands are at different levels, there is a difference of potential across the bands. (This may also be explained with the Fermi level which must remain flat throughout the device, but this explanation is beyond the scope of the course.) This means that the conduction and valence bands are higher in the p-type silicon than in the n-type silicon as shown in Figure 15.20a.

Figure 15.19 Types of diode

The energy difference between the conduction band on the p-type side and on the n-type side leads to a 'built-in' potential difference, in other words an electric field, which is present even when there is no applied

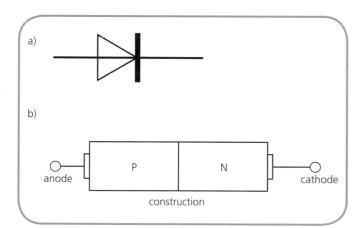

Figure 15.18 a) The circuit symbol for a diode. b) The layout of a p–n junction in a circuit

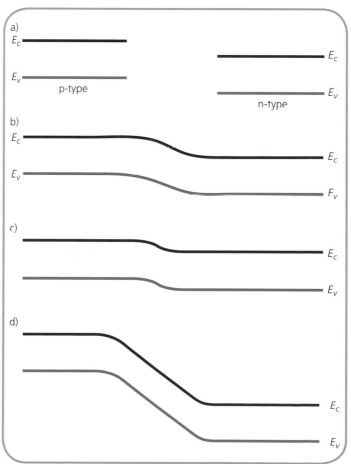

Figure 15.20 a) p- and n-type energy bands when separate. A p–n diode with b) no applied bias, c) applied forward bias and d) applied reverse bias. The built-in potential is always evident

potential difference. The region where the conduction and valence bands bend is called the **depletion layer**. This is where some of the extra electrons combine with the holes – or missing electrons – and reduce the effective doping of the silicon.

Now we get to what a diode can be used for. A diode essentially blocks current in one direction, but ensures that current does flow in the other. In Figure 15.20c we can see that by applying a forward bias (or a positive voltage) to a diode, we reduce the difference between energy bands, allowing current to flow far more easily. In Figure 15.20d we see the opposite when a reverse bias (negative voltage) is applied. Figure 15.21 shows

how this can be envisaged. If a wall is of moderate size, we need a bit of energy to get over it, but some of us can manage it. If the wall is made lower, it is a lot easier to get over and almost everyone can make it. If, however, the wall is made much taller, almost no one could climb over it. It is the same for electrons in diodes under these conditions. Figure 15.21 is an analogy for what is happening to the electric field in the p-n junction. Figure 15.21a shows the normal electric field, Figure 15.21b shows the electric field which is reduced because of forward bias and Figure 15.21c shows the increased electric field as a result of reverse bias.

A graph of current against voltage in a diode is shown in Figure 15.22. For silicon-based diodes, the large current will begin to flow at around 0.7 V of applied forward bias as this negates the built-in bias; this allows the moving charge to overcome the electric field. If a large negative bias is applied, diodes will break down. This can be a desirable effect. Some diodes are specifically designed to break down at a certain voltage, allowing them to be used as voltage sensors.

Figure 15.21 It can help to visualise diodes using an analogy of a wall. a) When there is no applied bias there is a wall, but with a little bit of energy, you (or the electron) can get over it. Not every electron will manage the journey. b) When there is applied bias, it is a little like the wall has been knocked down; it is so small that you can almost walk across it. c) Finally, if a reverse bias is applied, it is almost impossible to get across; the wall is just too tall

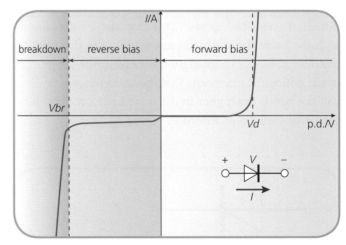

Figure 15.22 Current versus voltage for a diode. No values are shown as the values will depend on the individual diode. The current goes from about 0 to maximum almost as soon as the voltage reaches the value required to overcome the built-in bias (V_d). Very little current flows in the opposite direction until breakdown is reached (V_{br}). At this point, most normal diodes will be damaged and their characteristics will be permanently changed

As we can see, the most important section of the diode is the depletion layer. This is where the major changes take place and it is also the section which is altered in a basic diode to create other devices. There are multiple types of p–n junction-based devices and diodes. Some are used to ensure that current is constant while others are used to ensure that an exact voltage is maintained. Diodes are particularly useful in changing an alternating current to a direct current.

Earlier, we described the temperature dependence of the resistivity of semiconductors. It is possible to manufacture p–n junctions to ensure accurate temperature measurements. More complex structures can be created based on the same idea to make complicated devices such as transistors. Over the next few sections we will investigate two major practical uses of p–n junctions.

Light-emitting diodes

Diodes can be designed specifically to emit light. Such **light-emitting diodes** (LEDs) are already widely available and are used in a vast array of applications, from mobile phones to televisions. These diodes tend to be made mainly from GaAs (gallium arsenide) which is a III–V semiconductor. When electrons travel across the depletion region, they come together with holes. When this happens, they drop to a lower energy level and as energy is conserved, the energy lost by the electron–hole pair is released as a photon. The photon is emitted when the electron falls from the conduction to the valence band on either side of the junction.

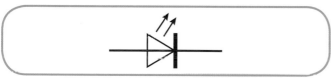

Figure 15.23 The circuit symbol for a light-emitting diode

The basic GaAs is altered with other semiconductor materials and elements to produce specific wavelengths of light. Red LEDs were the first to be mass produced since these can be made by the addition of phosphorus which was already being used in the manufacturing process. If nitrogen is added with the phosphorus then the wavelength of light is reduced leading to orange and yellow light. If no arsenic is used in the manufacturing process, green light can be produced.

Since the 1960s when LEDs were first mass produced, there has been an incredible amount of research into producing more efficient LEDs with a wider range of colours. As blue is one of the three primary colours of light, it was very important to create an LED of this

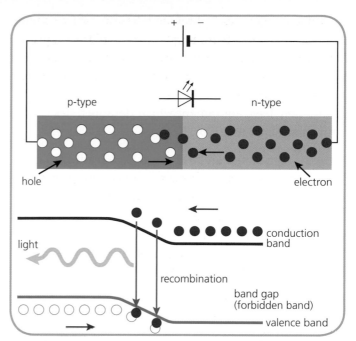

Figure 15.24 Energy bands in a light-emitting diode: the energy released as the electron drops from the conduction band to the valence band is emitted as light energy

colour, but the elements involved made it particularly difficult to manufacture. It took until the 1990s for these to become widely available. Some LEDs are created with wavelengths of light which are outside the visible range. Infrared LEDs are in most television remote controls. UV LEDs are also becoming more widespread.

The biggest difficulty comes when trying to create white LEDs. As you already know, white light is made up of lots of different wavelengths of light. In order to create a white LED, it is not possible to create photons of a single wavelength. There are two major methods of

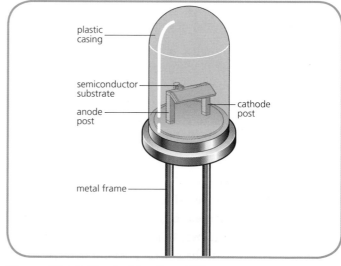

Figure 15.25 The inside of an LED

211

manufacturing white light. One relies on coating a blue LED to change the wavelength of some of the photons. The other is a method you can demonstrate yourself. If you mix the three primary colours of light (red, green and blue) together, you can produce white light. Your school may have equipment which demonstrates this with three separate LEDs.

If the voltage across an LED is too high leading to a large current in the junction, the diode may overheat. If this happens then the junction may be changed permanently. For this reason, we usually include a resistor in series with any LED in a circuit, unless we have other protection in the circuit to ensure that the current stays within allowable limits.

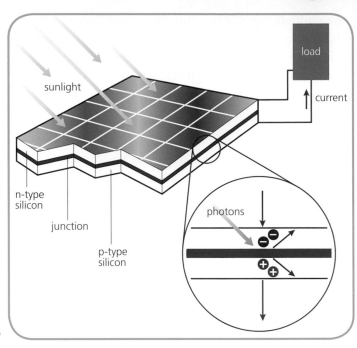

Figure 15.26 Individual solar cells are incredibly small, and we get almost no electrical energy from them

Activities

There are two experiments which can be used to investigate LEDs. First of all, because the semiconductor material is different, the built-in voltage is different. This means that the energy required to switch them on is different. Get a range of colours of LEDs and find out the voltage at which they switch on. Is the colour constant?

The next investigation regards the wavelength of light. Try to find out what material the LED is made from and what the strongest wavelength of light it produces is.

Research Task

LEDs are being used more and more in household lighting. What are the advantages and disadvantages of LEDs over traditional lighting methods? Is it possible to get the colour of light we are used to? Are LEDs capable of lighting a whole room?

LEDs are incredibly useful, but what if we need something stronger? Find out about the possibility of making lasers from p–n junctions. How do manufacturing techniques have to be altered? Are the same chemicals used to create the different colours? How is the light intensity increased?

Solar cells

Solar cells are also made using p–n junctions. While other semiconductor materials can be used, silicon tends to be the material of choice to make solar cells. The crystalline structure of the silicon is manufactured in a specific way to ensure the highest efficiency possible.

When a solar cell is exposed to light, photons enter the junction. When a photon is absorbed, electrons are raised from the valence band (leaving a hole behind) which creates a potential difference across the cell. This is the **photovoltaic effect** and causes a potential difference. This leads to a current flow. While a single photon will only lead to a tiny current, multiple photons release multiple electrons. Solar cells are rarely used individually; they are usually grouped together onto small solar panels. When grouped like this, a measurable current is able to flow and because it is a p–n junction, it can only flow in one direction. This means that the output of a solar cell is always a DC current.

Figure 15.27 When solar cells are arranged in panels like this one, we start to get useful amounts of energy

For Interest **The photovoltaic effect**

As we all know, the Sun gives off light and heat energy. We describe the light energy as particles called photons. In the early nineteenth century it was discovered by Antoine Henri Becquerel that light energy could create an electrical current in some materials, though at the time the idea of the photon had not yet been thought of and the reason for the effect was not fully understood. We now understand that the energy from the photon excites an electron, making it jump to a higher energy band. This creates a potential difference across the device and, in a p–n junction, means that the electrons can flow.

Timeline for the development of solar energy

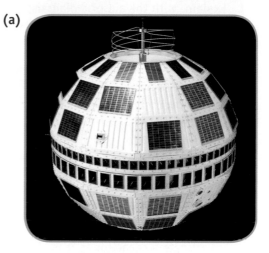
(a)

1839 The photovoltaic effect discovered by Becquerel

1883 First solar cell made by Charles Fritts

1921 Albert Einstein wins Nobel Prize for his work on the photoelectric effect

1954 Bell Labs in America creates the first modern solar cell

1962 Telstar satellite powered by solar cells

(b)

1967 Russia produces the first manned spacecraft with a section powered with solar panels called the Soyuz 1

1978 First solar-powered calculator

1987 First, World Solar Challenge, car race

2007 54-hour flight carried out by The Zephyr, a solar-powered plane made by QinetiQ

2009 Sharp announces the development of a solar-powered mobile phone

(c)

2010 Largest photovoltaic plant (the Sarnia Photovoltaic Power Plant) completed in Canada. The Sarnia plant has the highest peak output (nearly 100 MW) of any solar plant

Figure 15.28 a) Telstar satellite;
b) solar-powered calculator;
c) the Zephyr

Research Task

How can solar cells be used to produce energy for households? Is a solar panel a good investment or does further research need to be done? How does large-scale electricity production from solar cells compare to other renewable methods? What about non-renewable methods?

Questions

1 Explain, with the aid of a diagram, the difference between insulators and conductors in terms of energy bands.

2 Using your knowledge of physics, and with reference to energy bands, explain how and why the conductivity of a semiconductor depends on temperature.

3 Draw and annotate a diagram of a p–n junction. Indicate the energy levels and the movement of both holes and electrons. Show how this diagram would change in forward and reverse bias and describe the effect on the conduction of the semiconductor.

4 Design an experiment to investigate the frequency dependence of a solar cell, ensuring that you note down steps you would take to ensure a fair test.

Consolidation Questions

1 The following diagram shows a circuit set up to investigate a capacitor as it is charged.

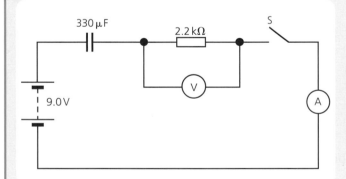

Figure 15.29

Initially, the capacitor is uncharged.

The capacitor has a capacitance of 330 µF and the resistor has a resistance of 2.2 kΩ. The battery, which can be assumed to have negligible internal resistance, has an EMF of 9.0 V.

a) i) Determine the initial current in the circuit when switch S is closed.
   ii) Determine the maximum energy which can be stored in the capacitor using this circuit.
   iii) Explain how the circuit could be changed to reduce the maximum energy stored in the capacitor.

b) A similar circuit is used to power the flash of a coloured strobe light.

The mean value of the frequency of photons is 4.8×10^{14} Hz.

On each cycle of the strobe light 3.82×10^{-3} J of energy are released.

Determine the number of photons released per cycle of the strobe light.

2 A battery for a specific circuit is designed to have an EMF of 4.5 V and an internal resistance of 1.5 Ω.

a) State what is meant by an EMF of 4.5 V.

b) The circuit to be used includes two resistors in series with the supply, as shown in Figure 15.30. The reading on the ammeter is 150 mA.

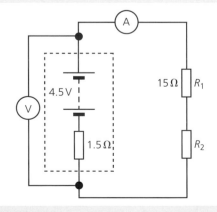

Figure 15.30

i) Calculate the resistance of R_2.

ii) Determine the reading on the voltmeter.

c) The same battery is used in the circuit shown in Figure 15.31. State what the effect of closing switch S will be in terms of the reading on the voltmeter. Explain your answer.

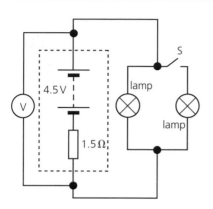

Figure 15.31

3 a) State the effect of doping on the resistivity of a pure semiconducting material such as silicon.

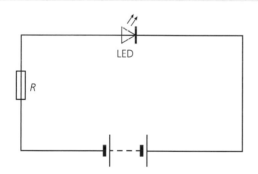

Figure 15.32

b) Figure 15.32 shows a circuit which includes a light emitting diode (LED). The LED is an example of a p-n junction.

i) With reference to energy, state how the LED emits photons. You must explain your answer.

ii) The light from the LED is shone through a grating, as shown in Figure 15.33. This LED produces monochromatic light with a wavelength of 480 nm. Determine the spacing of the grating.

4 A heating element is used in a portable towel warmer. The towel warmer operates with two cells each with

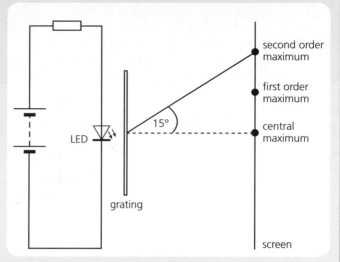

Figure 15.33

an EMF of 1.5 V. The internal resistance of each cell is 0.40 Ω. The resistance of the heating element is 4.8 Ω. The circuit is shown in Figure 15.34.

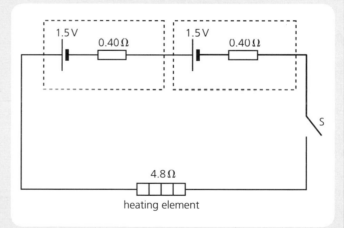

Figure 15.34

a) Calculate the total resistance of the circuit when switch S is closed.

b) Determine the current in the heating element when the device is on.

c) Determine the maximum energy transformed per second by the heating element.

d) Over time, it is found that the internal resistance of these particular cells decreases. Determine the effect of this on your answer to part **c)**. You must justify your answer.

Appendix 1: Assessment – Question paper and Assignment skills

This section is intended to give an indication of the skills you will gain by doing the Higher Physics course. You will gain much knowledge of the physics of everyday life and some more unusual occurrences but you will also be able to develop your skills in such areas as data handling, checking evidence, presenting information and critical evaluation.

What you need to do in order to pass the course is given in the National Course Specification which is published by the Scottish Qualifications Authority (SQA). Details of the specification are given here for your information.

Assessment skills

The course assessment meets the key purposes and aims of the course by assessing breadth, challenge and application.

Candidates will apply breadth and depth of skills, knowledge and understanding from all areas of the course. They will demonstrate skills of scientific inquiry, using their knowledge of physics to carry out an extended investigation in physics and to present their findings.

The assessment has three components: two question papers and an assignment which combine to ensure full coverage of the course.

Question paper 1: Multiple choice questions, 25 marks total; 45 minutes duration

Question paper 2: It contains restricted-response and extended-response questions. The paper has a total of 130 marks; 2 hours and 15 minutes duration

Assignment: Nominally 8 hours of classroom-based experiment and research of which 2 hours is allowed for the writing of the report. It is worth 20 marks.

These marks are combined to give a numerical score for each candidate and this score forms the basis for their grade.

The marks for each component are scaled. This means they are altered to increase or decrease their relevant impact.

Paper 1   25 marks scaled to 25

Paper 2   130 marks scaled to 95

Assignment   20 marks scaled to 30

This gives a maximum total of 150. This is converted to a percentage and the grade awarded to the candidate is based upon that.

For example, a candidate receives 18 marks for paper 1, 78 marks for paper 2 and 16 for their assignment.

This would be scaled to $18 + 57 + 24 = 99$

$$99/150 = 66\%$$

This would normally lead to an award of a grade B.

Assignment skills
Summary

During the assignment, candidates will develop the key skills of scientific inquiry and related physics knowledge and understanding. It allows assessment of skills that cannot be assessed through the question paper, for example the handling and processing of data gathered from experimental work by the candidate.

Assignment overview

The assignment gives candidates an opportunity to demonstrate the following skills, knowledge and understanding:

- applying physics knowledge to new situations, interpreting information and solving problems

- planning and designing experiments/practical investigations to test given hypotheses or to illustrate particular effects

- recording detailed observations and collecting data from experiments/practical investigations

- selecting information from a variety of sources

- presenting information appropriately in a variety of forms

- processing information (using calculations, significant figures and units, where appropriate)

- drawing valid conclusions and giving explanations supported by evidence/justification

- quantifying sources of uncertainty

- evaluating experimental procedures and suggesting improvements

- communicating findings/information effectively.

The assignment has two stages:

- research

- report.

The research stage must involve experimental work which allows measurements to be made. Candidates must also gather data/information from the internet, books or journals.

Candidates must produce a report on their research, presenting the aim, results and conclusions of their experimental work.

The SQA recommend that no more than 8 hours is spent on the whole assignment. A maximum of 2 hours is allowed for the report stage.

Research stage

Candidates research and report on a topic that allows them to apply skills and knowledge in physics at a level appropriate to Higher. The topic must be chosen with guidance from teachers and/or lecturers and must involve experimental work. It will be related to something you have covered in your studies, but it could be in an unfamiliar context. You will have to research the physics behind the topic. Internet searching is preferred by most people these days as the various engines trawl through all the available sites or articles, but you need to be careful about the validity of some sites. There are many sites which may seem scientific but which are not as impressive or as accurate

as they first seem. Check whether the site has a contact address. This is a good sign. The URL can also give an indication of a website's authority. For example, if the URL ends in ac.uk, this means it is an accredited university or college and would generally be reliable. It is better if you obtain evidence from a range of sources and this can be used to support your point.

Using more than one source may also give you a better perspective on the topic and if you can be critical in your use of the information, it will provide evidence of the skills the section is hoping to assess.

The following section outlines ways in which you can gather and use any data required for the assignment. It shows how you could present it and also analyse it to ensure you have confidence in your conclusions. Not all topics will allow you to do all the activities in the section but you should be able to take appropriate 'parts' of the section and use them in your own report.

Assignment experimental work
1 Data collection

The skills examined in the assignment include the ability to record and analyse experimental data from a scientific experiment. In the following example an experiment to determine the relationship between the length of a pendulum and its period is used to demonstrate how data might be collected and analysed.

When attempting to determine the relationship between two variables the following terms are used:

- **Independent variable:** the quantity which is varied by the experimenter. In mechanics experiments, this is very often the time of measurement. In this case, we will change the length of the pendulum.

- **Dependent variable:** the quantity which changes as a result of a change in the independent variable. In this case, we will measure how the pendulum period (the time for one complete swing of the pendulum) is altered by changing the pendulum length.

- **Control variables:** these are all the other variables which might have an impact on the dependent variable but which we attempt to keep constant so as to determine the true relationship between independent and dependent variables. In this case we will control the mass of the pendulum bob, use the same timer each time and maintain a constant laboratory temperature.

It is important to use as wide a range of independent variable values as possible to ensure the relationship determined may be applied. Equally, a reasonable number of values of independent variable should be used within this range to ensure the correct shape of the relationship may be derived.

For each value of independent variable, the measurement of the dependent variable should be repeated a number of times and averaged in order to both confirm the accuracy of the measured value and improve the precision associated with the average. Typical minimum data collection requirements would be to measure a wide range of five values of independent variable and repeat the dependent variable measurement five times.

In our example of the length of a pendulum and period, this gives a data table which looks like Table AP1.1 below.

Notice that the data table must have column labels and units to identify what is being recorded in each cell.

In planning the experiment, it is good practice to design the data table in advance of carrying out the work in order to clarify your thinking. This type of data table lends itself well to using a spreadsheet program like Microsoft Excel® for the calculation of the average and subsequent analysis.

When recording the data it is important that each value is entered to the same precision – in other words, the same number of decimal places – within any one column. We will deal with uncertainty analysis later in this section.

In our example, we might then expect a series of data which looks like Table AP1.2.

Notice that the times we are recording here are relatively short – between about 1 and 2 seconds. We might improve the accuracy and precision of the time variable by measuring the time for, say, 10 swings and then dividing by 10 to derive the period. In this example, let us proceed with the data as shown in Table AP1.2.

Dependent variable

Pendulum length/m	Period 1/s	Period 2/s	Period 3/s	Period 4/s	Period 5/s	Average period/s

Independent variable

Table AP1.1 Example data table layout

Pendulum length/m	Period 1/s	Period 2/s	Period 3/s	Period 4/s	Period 5/s	Average period/s
0.202	1.07	0.99	1.00	1.06	1.05	1.03
0.350	1.29	1.36	1.42	1.24	1.29	1.32
0.513	1.51	1.60	1.56	1.68	1.59	1.59
0.688	1.89	1.80	1.74	1.81	1.84	1.82
0.914	2.05	1.94	1.91	1.97	1.88	1.95

Table AP1.2 Pendulum period vs length data

2 Uncertainty estimates

Any measured value is at best an estimate of the true value of the measurement. The uncertainty in measurement arises from a number of sources:

- Scale-reading uncertainty is a measure of how well an instrument scale can be read.

- Random uncertainties occur when an experiment is repeated and slight variations occur. These may be reduced by taking repeated measurements.

- Systematic uncertainties occur when the experimental process or measurements are poorly designed. These tend to shift the readings taken so that they are either all too small or all too large.

By definition, the scale-reading uncertainty is estimated as one whole division of the scale if the measurement is being made on an electronic digital meter. If the measurement is made using an analogue meter (a meter with a moving needle or other fixed scale), the systematic uncertainty is estimated as a half division of the scale.

In the pendulum example, the length is measured with a wooden metre stick ruled off in millimetre intervals and the timer has an electronic display which reads to 0.01 seconds.

Therefore, the scale-uncertainty estimates in our example are:

- lengths ± 0.0005 m (analogue scale)

- times ± 0.01 s (digital scale).

The random uncertainty in a repeated measurement of the dependent variable may be estimated using the formula:

$$\text{random uncertainty} = (\text{max. value} - \text{min. value})/\text{number of values}$$

The scale- and random uncertainties above are known as the absolute uncertainty values as they have the same units as the measurements to which they refer. A percentage uncertainty may also be calculated using the formula:

$$\text{percentage uncertainty} = (\text{absolute uncertainty}/\text{measurement value}) \times 100$$

Where more than one uncertainty exists in a measurement, the largest percentage uncertainty should be used as an estimate of the actual percentage uncertainty in the results.

In the pendulum example, then, the uncertainty estimates for each row are shown in Table AP1.3.

Random		Scale-reading	
ΔAverage period/s	ΔAverage period/%	ΔAverage period/s	ΔAverage period/%
0.02	2%	0.01	1%
0.04	3%	0.01	1%
0.03	2%	0.01	1%
0.03	2%	0.01	1%
0.03	2%	0.01	0.5%

Scale-reading	
ΔLength/m	ΔLength/%
0.0005	0.2%
0.0005	0.1%
0.0005	0.1%
0.0005	0.1%
0.0005	0.1%

Table AP1.3 Uncertainty estimates

Note that the Greek letter Δ may be used to indicate the uncertainty in the value as shown in Table AP1.3.

Given the above, the %Δ in the average period should be estimated using the %Δ values from the random uncertainty calculations as these are the largest percentage values in each case.

3 Expressing values and uncertainties

Any reading or value derived from an experiment should then be written as the value ± the associated uncertainty, for example time = (0.28 ± 0.03) s.

Note that:

- both numbers are written inside the brackets and so have the same unit (seconds in this example)

- the reading is expressed to the same precision as the uncertainty (two decimal places in the example)

- the uncertainty is written with only one significant figure (if we are uncertain in the second decimal place then greater accuracy is not possible).

The uncertainty in the value may also be expressed as a percentage. In this case above, time = 0.28 s ± 10%

(Actually, % uncertainty = 0.03/0.28 × 100 = 10.7% but as above, the uncertainty is always written with a single significant figure.)

4 Graphing data

When graphing data it is good practice to use the absolute uncertainty in the value being plotted to add error bars to indicate the area within which the true value of the data point lies.

In our example the pendulum length uncertainties are so small compared to the x-axis graph scale that they cannot be seen or plotted by hand. The uncertainty in the time, however, does generate usable error bars and they are shown in Figure AP1.1.

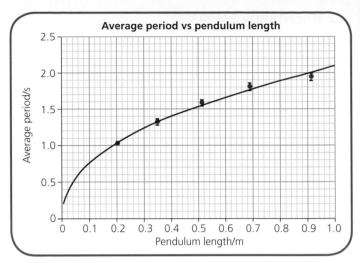

Figure AP1.1

It is clear from Figure AP1.1 that the relationship between the pendulum period and length is not directly proportional. In fact, the equation which governs this relationship is $T = 2\pi\sqrt{(l/g)}$

The various elements of this equation are detailed in Table AP1.4.

It is nearly always necessary for us to plot a graph of data which shows a straight-line relationship between the independent and dependent variable. This is what allows us to write equations of the form

$$A = B \times C$$

If you look at the sheet detailing the relationships required for Higher Physics, you will note that most of the equations we use are of this form. For the data and equation we have in our pendulum example, we need to process the data further to be able to plot a graph which is in the form of a straight-line relationship.

Any straight line may be represented by the equation

$$y = mx + c$$

where y is the dependent variable, x is the independent variable, m is the gradient of the line and c is the y-axis intercept.

Quantity	Name	Unit	Symbol
T	pendulum period	second	s
l	pendulum length	metre	m
g	acceleration due to gravity	metres per second²	$m\,s^{-2}$

Table AP1.4

For the pendulum then:

$$T = 2\pi\sqrt{(l/g)}$$
$$T^2 = 4\pi^2(l/g)$$
$$T^2 = (4\pi^2/g)l$$

This is now of the form $y = mx + c$

where $y = T^2$; $m = (4\pi^2/g)$; $x = l$ and $c = 0$

This means that if we plot T^2 versus l we should find a straight-line relationship which passes directly through the origin (as $c = 0$) and has a gradient $m = (4\pi^2/g)$.

So we should now be able to calculate the acceleration due to gravity acting on the pendulum directly from the gradient of the graph.

The data to be plotted are now as given in Table AP1.5.

Pendulum length/m	(Average period)^2/s^2
0.20	1.07
0.35	1.74
0.51	2.53
0.69	3.31
0.91	3.80

Table AP1.5

(The following is beyond the level of this course: where a value is calculated using a power relationship, the percentage uncertainty in the value is multiplied by the power to get the percentage uncertainty in the result. In this case as period is squared, the percentage uncertainty is doubled to calculate the size of the error bars in (average period)^2.)

The relationship is now graphed as shown in Figure AP1.2.

In Figure AP1.2, Excel® has been used to plot a linear trend line and display the equation of the line. Note that the line does not pass through the origin as expected. This suggests a systematic uncertainty in our experiment which has caused the data to be shifted – perhaps the length measurement included only the string length but not the radius of the bob or perhaps the timer was poorly calibrated.

The gradient is shown as $m = 3.9324$, but we know

$$m = (4\pi^2/g) \text{ so}$$

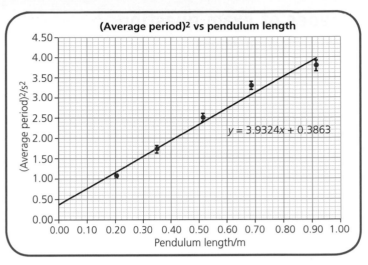

Figure AP1.2

$$(4\pi^2/g) = 3.9324$$
$$g = 4\pi^2/3.9324$$
$$g = 10.04 \, \text{m s}^{-2}$$

The uncertainty in the value for g derived from the graph must also be estimated by determining the uncertainty in the gradient value used in the calculation. In our example, we can estimate steepest and shallowest best fit lines and then use their gradients in determining maximum and minimum values of g.

Alternatively, we may use Excel® to estimate the absolute uncertainty in the gradient using the built-in line statistics function LINEST which returns the values of both gradient and intercept and their associated absolute uncertainties.

In our example, this gives:

$$m = 3.9$$
$$\Delta m = 0.3$$
$$c = 0.4$$
$$\Delta c = 0.2$$

Therefore the percentage uncertainty in m (i.e. %Δm) is given by

$$\%\Delta m = 0.3/3.9 \times 100 = 8\%$$

This is taken to be the percentage uncertainty in our value of g and so our experimental evaluation gives

$$g = 10.12 \, \text{m s}^{-2} \pm 8\%$$

or

$$g = (10.1 \pm 0.8) \, \text{m s}^{-2}$$

Types of exam question

Numerical questions

Many of the questions in an exam paper will be calculations. It is essential that these are laid out carefully as this will make them easier to follow and ensure that you gain as many marks as possible.

There are four stages in answering a two-mark numerical question:

1 Identify the variable you want to calculate and write down information from the question.

2 Select and write down an equation.

3 Substitute values into the equation.

4 Rearrange and calculate the answer.

It is important that we highlight the final answer as this makes it clear to the marker that this is the end of the question. The answer is given to the same accuracy (significant figures) as the variables in the question. This principle remains even with more difficult examples.

Example 1

A basic example is shown here.

Calculate the reading on the ammeter in Figure AP1.3.

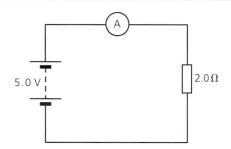

Figure AP1.3

Step 1

We want to calculate the current, I, and we are given details of V and R. So we write:

$I = ?$
$V = 5.0\,V$
$R = 2.0\,\Omega$

Step 2

We look at the three variables available – V, I and R – and select Ohm's Law, $V = IR$.

Step 3

Substitute the known values into the equation.

$5.0 = I \times 2.0$

Step 4

Rearrange and solve the equation.

$I = 5.0/2.0$
$\underline{I = 2.5\,A}$

Example 2

A 50.0 kg satellite is placed in orbit around a planet of mass 6.45×10^{18} kg. How close must it be for the force between the satellite and planet to be 6.20 N?

Step 1

We want to calculate the distance and in the case of a planet this will be a radius.

Step 2

We need an extra detail, though, because the gravitational field constant, G, is needed. You will find this in the data sheet at the front of your exam paper. This needs to be added to our list of variables.

$r = ?$
$m_1 = 50.0\,kg$
$m_2 = 6.45 \times 10^{18}\,kg$
$F = 6.20\,N$
$G = 6.67 \times 10^{-11}\,m^3\,kg^{-1}\,s^{-2}$
$F = \dfrac{Gm_1m_2}{r^2}$

Step 3

Substitute the values.

$6.20 = 6.67 \times 10^{-11} \times (50.0 \times 6.45 \times 10^{18})/r^2$

Step 4

Rearrange and solve the equation.

$6.20 \times r^2 = 6.67 \times 10^{-11} \times (3.225 \times 10^{20})$
$r^2 = 6.67 \times 10^{-11} \times (3.225 \times 10^{20})/6.20$
$r^2 = 3.469 \times 10^9$
$r = \sqrt{(3.469 \times 10^9)}$
$\underline{r = 58.9\,km}$

Some numerical questions will give you the solution, but ask you to show how this can be reached. It is essential in these instances that you do not use the solution at the start of your answer. Treat it like any other numerical question.

Example 3

Show that an 85 kg man moving at $3.0\,ms^{-1}$ has 380 J of kinetic energy.

$E = ?$
$m = 85\,kg$
$v = 3.0\,ms^{-1}$

$E = \tfrac{1}{2}mv^2$
$E = 0.5 \times 85 \times 3.0^2$
$\underline{E = 382.5\,J = 380\,J}$

A question may include a diagram or a graph from which you should take your data. This, once again, should follow the same pattern, although questions may require two stages.

Example 4

What is the critical angle of the glass block in the diagram below?

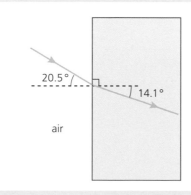

Figure AP1.4

The details in the diagram allow you to calculate the refractive index. Using the refractive index you can calculate the critical angle. This means we must split the question into two parts.

$\theta_1 = 20.5°$
$\theta_2 = 14.1°$
$n = ?$

$n = \sin\theta_1/\sin\theta_2$
$n = \sin 20.5°/\sin 14.1°$
$n = 1.44$

We then use these data to go through the process again.

$\theta_c = ?$
$n = 1.44$

$\sin\theta_c = 1/n$
$\sin\theta_c = 1/1.44$
$\underline{\theta_c = 44.0°}$

Open-ended questions

Open-ended questions are a new addition to this revision of the Higher Physics course. They are used to give you the opportunity to show a wider knowledge of a subject, but also to explain what you understand. As with the numerical questions, there is a method for working through these.

1 As with every question, read it very carefully. Some of the physics in the question may not be correct and you should be ready to state that.

2 Decide what you need to answer; select a section or sections of the question to focus on.

3 Decide what you think the correct physics is; you may agree or disagree with what is written.

4 Plan (you may wish to write down a brief plan) an answer thinking about:

- the variables in the question

- which law(s) of physics you should mention

- what you want to agree and disagree with

- a diagram which may be useful to demonstrate your understanding.

5 Write your answer, including all information which is relevant.

6 Read your work and make sure you have given a complete answer.

There is rarely one 'perfect' answer to open-ended questions. They will be marked on individual merits. Similarly, the length of the answer is not necessarily important, but be aware that you are unlikely to cover the physics involved in enough detail in less than a paragraph.

It is possible to score 0, 1, 2 or 3 marks for an open-ended question. To attain the full three marks you must:

- demonstrate full comprehension of the problem

- relate the problem to physical laws and/or equations

- show an overall in-depth understanding of the physics related to the situation.

Two marks will be awarded for:

- a basic understanding of the problem

- reference to relevant physical principles.

One mark will be awarded in situations which demonstrate:

- a limited understanding of the problem

- some statement in the response which relates a physical principle to the situation.

While the answer is not negatively marked, you will be given no marks for:

- irrelevant physical principles

- restating parts of the question

- incorrect statements of related physics.

As with all descriptive questions, it is essential that you use correct terminology and that the physics you state is relevant to the situation which is being described. On the next pages are some examples with sample answers. They are not the only answers, but give an indication of good answers.

Example 1

A student performs an experiment to investigate the effect of internal resistance using this circuit. She alters the resistance of the variable resistor and measures the potential across it as the resistance increases.

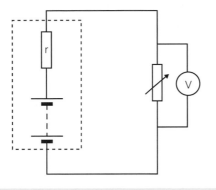

Figure AP1.5

The student writes as a conclusion:

'As the resistance of the variable resistor changed, the potential across it increased. This is because the internal resistance of the supply is getting smaller. This means that less current is used there, and more is in the potentiometer.'

Comment on the student's conclusions.

Suggested answer:

The student has misinterpreted the results. The reading on the voltmeter, 'the potential difference', will increase as the resistance of the variable resistor increases. This is because the circuit acts in a similar way to a potential divider. The internal resistance does not change but the potential difference across the internal resistance gets smaller. This is because the current in the circuit decreases as the total resistance increases (due to Ohm's Law, $V = IR$). As the internal resistance remains the same, V in the equation $V = Vs \times R/(R + r)$ becomes larger. Finally, the last sentence suggests that current is 'used' in the internal resistance; this is not the case. The current is the same at all points in a series circuit.

Example 2

A teacher has two materials, A and B. He knows that one is a conductor and one is a semiconductor so he measures the resistance after the materials have been in the freezer and at room temperature. He measures the resistance of each block once at each temperature. The resistance of A is higher at room temperature and B is lower, when compared to the freezer results. The teacher declares 'B is the semiconductor'.

Discuss this experiment and the teacher's conclusion using your knowledge of physics.

Suggested answer:

The teacher is performing an experiment based on the idea that the conductivity of a semiconductor will increase as temperature increases. This is not the case for conductors, so he is able to judge the two materials based on their resistivity. As the temperature increases, the resistivity of material B decreases and this dependence is what we would expect to see in a semiconductor. The experiment could be improved by increasing the temperature difference between the two measurements and by repeating the experiment to make it more reliable. Despite this, his conclusion based on the evidence is correct.

Example 3

In car crash tests, dummies are seen to move forward and sometimes travel out of the front window of cars. Explain in terms of your knowledge of motion why the dummies do so.

Suggested answer:

Objects in motion will continue at that velocity unless acted on by another force. When the cars crash and stop, the dummies continue at their initial velocity until acted on by another force. This force could be provided by the seat belt or airbag or, if they are not restrained, the windscreen. To outside observers, the dummies appear to be coming out of the car. They are doing so because the car has stopped suddenly.

Example 4

Two students were discussing what would happen if they were in a lift that was falling towards the ground. One said, 'If you jump up in the air just before the floor of the lift makes contact with the ground, you would be saved'. The other said that the jump would make little or no difference to your safety.

Who is correct?

Suggested answer:

As the lift falls towards the ground it will be travelling fairly quickly (possibly around $15\,\text{m s}^{-1}$). If you jump upwards at the point of impact at, say, $2\,\text{m s}^{-1}$, this will have the effect of making your resultant velocity only $13\,\text{m s}^{-1}$ downwards. This may reduce the impact, but not by much. (And the roof may come down on top of you anyway!) There are inherent difficulties with timing your jump given that you may not see the ground rushing towards you.

Exam technique

There are a few points to remember when sitting exams:

1 You do not have to answer the paper in order. Work through the questions in the best order for you. Go back to questions you find more difficult later.

2 Write carefully – the marker needs to be able to read the answers.

3 Answer all of the questions – especially in the multiple choice section!

4 Sometimes a question will build on one of your previous answers. It is OK to use your answer – even if you are not sure whether it is correct.

5 Look at how many marks a question is worth. This will help to guide your answer, especially in explain/describe questions.

Appendix 2: Useful formulae

Formula
$v = u + at$
$s = ut + \frac{1}{2}at^2$
$v^2 = u^2 + 2as$
$s = \frac{1}{2}(u+v)t$
$F = ma$
$E_w = Fd$
$E_p = mgh$
$E_k = \frac{1}{2}mv^2$
$P = \frac{E}{t}$
$p = mv$
$Ft = mv - mu$
$d = \bar{v}t$
$s = \bar{v}t$
$W = mg$
$E = hf$
$E_w = QV \qquad (W = QV)$
$E = V + Ir$
$Q = It$
$R_T = R_1 + R_2 + \ldots$

Formula
$\frac{1}{R_T} = \frac{1}{R_1} + \frac{1}{R_2} + \ldots$
$V_1 = \left(\frac{R_1}{R_1 + R_2} \right) V_S$
$V = IR$
$P = IV = I^2 R = \frac{V^2}{R}$
$V_{peak} = \sqrt{2} V_{rms}$
$C = \frac{Q}{V}$
$E = \frac{1}{2}QV = \frac{1}{2}CV_2 = \frac{1}{2}\frac{Q^2}{C}$
$\frac{V_1}{V_2} = -\frac{R_1}{R_2}$
$T = \frac{1}{f}$
$v = f\lambda$
path difference $= m\lambda$ or $(m + \frac{1}{2})\lambda$ where $m = 0, 1, 2 \ldots$
$d \sin\theta = m\lambda$
$n = \frac{\sin\theta_1}{\sin\theta_2}$
$\frac{\sin\theta_1}{\sin\theta_2} = \frac{\lambda_1}{\lambda_2} = \frac{v_1}{v_2}$
$\sin\theta_c = \frac{1}{n}$

Formula
$I = \dfrac{k}{d^2}$
$E_2 - E_1 = hf$
$I = \dfrac{P}{A}$
$E_k = hf - hf_0$
$F = G\dfrac{m_1 m_2}{r^2}$
$t' = \dfrac{t}{\sqrt{1 - \left(\dfrac{v}{c}\right)^2}}$
$l' = l\sqrt{1 - \left(\dfrac{v}{c}\right)^2}$

Formula
$f_o = f_s\left(\dfrac{v}{v \pm v_s}\right)$
$z = \dfrac{\lambda_{\text{observed}} - \lambda_{\text{rest}}}{\lambda_{\text{rest}}}$
$z = \dfrac{v}{c} \quad v = H_0 d$
$E = mc^2$
$\text{random uncertainty} = \dfrac{\text{max. value - min. value}}{\text{number of values}}$

Answers to in-text questions

Chapter 1: Motion

1 a) Distance = 880 m; magnitude of displacement = 0 m

b) Distance = 1400 m; magnitude of displacement = 1000 m

c) Distance = 55 m; magnitude of displacement = 25 m

d) Distance = 2 260 800 km; magnitude of displacement = 0 km

e) Distance = 20 km; magnitude of displacement = 18.54 km

f) Distance = 2.8 m; magnitude of displacement = 0.4 m

2 a) Blue graph = 13 m s^{-1}; Red graph = 10 m s^{-1}

b) 6 s, 24 s

3 a) 7 m s^{-1} **b)** 4.5 m s^{-1} **c)** 3.6 s, 10.8 s

4 a) A   **b)** The slope of the graph was greater in A than B.

5 a)

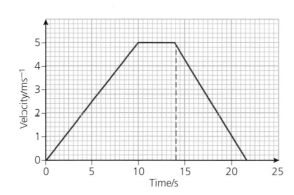

b)

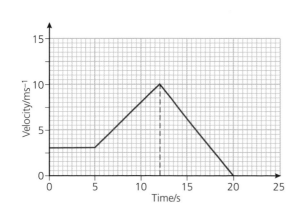

c)

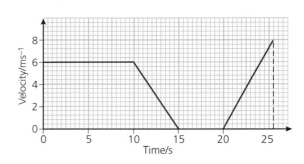

6 a) 3.0 s   **b)** 29.4 m s^{-1}

7 a) 122.63 m   **b)** 49.05 m s^{-1}

8 a) 0.39 m s^{-2} **b)** 6318 m

9 8.4 m s^{-2}

10 a) 2.37 m s^{-2} **b)** 6.16 m s^{-1}

11 a) 16 m s^{-1} **b)** 32 m   **c)** 7.07 s

12 a) 96 m   **b)** -0.5 m s^{-2}

13 a) 0.3 m s^{-2} **b)** 190 m

14 a) 10 m   **b)** 41 m   **c)** 5.1 m s^{-1}
d) -0.75 m s^{-2}; 0 m s^{-2}; -2.5 m s^{-2}

15 a) -0.667 m s^{-2} **b)** -0.375 m s^{-2} **c)** 2 m
d) No, downhill displacement was greater
e) -0.25 m s^{-1}

16 Diver leaves aircraft and accelerates until reaching terminal velocity. Velocity slows dramatically as parachute opens. Velocity drops to zero as diver lands.

17 Variable answers. They may maintain an average speed, but within the cameras cars may reach high speeds.

18

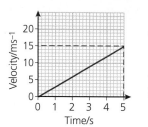

19

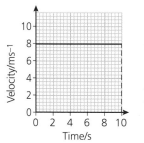

20

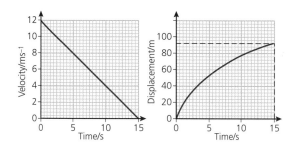

Chapter 2: Forces, energy and power

1 **a)** 10.15 N **b)** 2.5 m s^{-2}

2 **a)** 6 N **b)** 17.8° **c)** No difference

3 **a)** 1.7 m s^{-2} **b)** 2.6 m s^{-1}

4

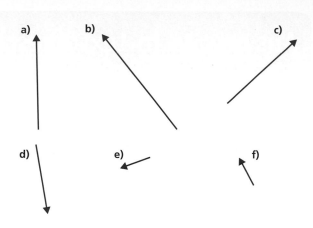

5 **a)** 11.67 N; 31° **b)** 25 N; 59° **c)** 12.65 N; 71.6°
d) 8.6 N; 35.5° **e)** 250 N; 36.9° **f)** 18.5 N; 22.5°
g) 17.8 N; 82.5° **h)** horizontal; 188° **i)** 500 N; vertical
j) 26 N; vertical

6 **a)** 3.4 m s^{-1} **b)** 120 m

7 **a)** 180 m s^{-1} **b)** 240 m s^{-1}

8 Due to jet stream rotation of the Earth

9 **a)** $E_p = mgh = 92 \times 9.8 \times 900 = 811\,440\,\text{J}$

b) $E_k = \tfrac{1}{2}mv^2 = \tfrac{1}{2} \times 92 \times 35^2 = 56\,350\,\text{J}$

c) A lot of energy has been transformed to heat as the skis carve through the snow and the skier races through the air.

10 **a)** 2.94 J **b)** 2 m s^{-1} **c)** 1.5 J and 1.5 J

11 **a)** 78 480 J **b)** 9000 J **c)** 44.3 m s^{-1}

12 **a)** 29.4 J **b)** 2.4 × 10^{-5} J

Chapter 3: Collisions, momentum and energy

1 **a)** 20 kg m s^{-1} and 0 kg m s^{-1} **b)** 20 kg m s^{-1}
c) 2.86 m s^{-1}

2 **a)** 6.7 m s^{-1} **b)** 2 m s^{-1} **c)** 2 m s^{-1}
d) 8.3 m s^{-1} **e)** −2 m s^{-1}(left) **f)** 0 m s^{-1}

3 **a)** 700 kg m s^{-1} **b)** 6.5 m s^{-1}

4 **a)** 12 000 kg m s^{-1} **b)** 36 000 J **c)** 2.4 m s^{-1}
d) 14 400 J **e)** The energy is transformed in deforming the metal, as sound, etc.

5 **a)** 2 m s^{-1} **b)** Before 1.8 J; after 1.8 J
c) Sound and heat

6 **a)** 10 kg m s^{-1} **b)** 0.05 m s^{-1} (90 kg player's direction)
c) 2004.75 J

7   $16.8\,\mathrm{m\,s^{-1}}$ to the left

8   **a)** $7.67\,\mathrm{m\,s^{-1}}$ **b)** $1.1\,\mathrm{m\,s^{-1}}$ **c)** $2117.5\,\mathrm{J}$

9   **a)** $2.4\,\mathrm{m\,s^{-1}}$ to the left   **b)** $2400\,\mathrm{N}$

10 **a)** $2.25\,\mathrm{kg\,m\,s^{-1}}$ **b)** $450\,\mathrm{N}$
    c) It slows during contact with the ball.

Chapter 4: Projectiles and satellites

1   **a)** $78.5\,\mathrm{m}$ **b)** $39.2\,\mathrm{m\,s^{-1}}$

2   **a)** $0.55\,\mathrm{s}$ **b)** $5.4\,\mathrm{m\,s^{-1}}$ **c)** $1.1\,\mathrm{m}$
    d) $20°$ from vertical

3   **a)** $2.26\,\mathrm{s}$ **b)** $180\,\mathrm{m}$ **c)** $15.5°$ from vertical

4   **a)** Horizontal $= 1.8\,\mathrm{m\,s^{-1}}$; vertical $= 0.85\,\mathrm{m\,s^{-1}}$
    b) $0.5\,\mathrm{s}$ **c)** $0.95\,\mathrm{m}$

Chapter 8: Electric fields

1   **a)** See Figure 8.3 on page 84.

    b) See Figure 8.6 on page 85.

    c) See Figure 8.4 on page 85.

2   The field around the positive charge would be stronger, meaning that the field lines would be closer together.

3   Both particles would move from the positive plate to the negative plate in a straight line. The particle with $+2q$ would require double the work done to reach this position, so would have double the kinetic energy when it reaches the negative plate. As a result, it would move more quickly from the positive to negative plate.

4   $E_w = 4.0 \times 10^{-17}\,\mathrm{J}$

5   $V = 6\,\mathrm{kV}$

6   **a)** $2.4 \times 10^{-16}\,\mathrm{J}$

    b) $2.3 \times 10^7\,\mathrm{m\,s^{-1}}$

7   **a)** $A = 1.5 \times 10^8\,\mathrm{m\,s^{-1}}$; $B = 2.1 \times 10^8\,\mathrm{m\,s^{-1}}$;
    $C = 2.6 \times 10^8\,\mathrm{m\,s^{-1}}$; $D = 3.0 \times 10^8\,\mathrm{m\,s^{-1}}$

    b) After half way, the electron is travelling at more than two thirds of the speed of light. Relativistic effects cause effective mass to increase by more than 10% at half of the speed of light, and so the effective mass of the electron is rapidly increasing after this point. By point C, the effective mass of the electron will have doubled.

8   $530\,\mathrm{MV}$, assuming that relativistic effects do not occur

9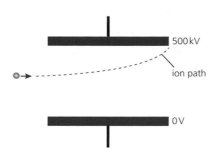

10 1 joule per coulomb

11 $1.44 \times 10^{-16}\,\mathrm{J}$

12 **a)** $4.25 \times 10^{-11}\,\mathrm{J}$

    b) $4.8 \times 10^{-19}\,\mathrm{C}$

    c) 3 protons and 2 neutrons based on mass and charge

13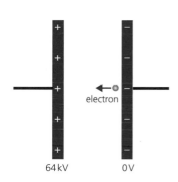

    Energy required is $10^{-14}\,\mathrm{J}$ so parallel plates with a p.d. of $64\,\mathrm{kV}$ will accelerate the electron to this velocity.

14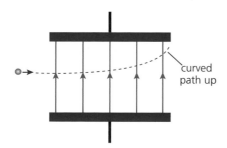

15

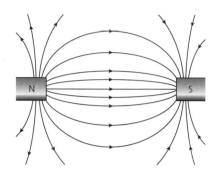

16

a) See Figure 8.23 on page 93.

b)

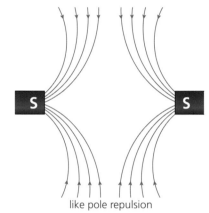

like pole repulsion

c) See Figure 8.24 on page 93.

17 Multiple correct answers

18 B to A

19 a) Out of page **b)** From top to bottom
  c) Out of page (though the electron moves into the page) **d)** From right to left (though the electron goes the other way)

20 a) Out of page **b)** From top to bottom
  c) From top to bottom **d)** From left to right

21 a) From top to bottom **b)** From bottom to top
  c) Out of the page **d)** Into the page

22 a) Positive **b)** Negative **c)** Positive **d)** Positive

23 One field to control left-right, another to control up-down

24 The proton has more mass.

25 a) The energy required to accelerate the particle will increase as the velocity increases, and the frequency of the potential must match the movement of the particle. **b)** As the particle approaches the speed of light **c)** Multiple correct answers – any valid experimental description.

26 The tubes become smaller as the particle accelerates in order to keep the acceleration constant without changing the frequency of the changing electric field.

27 Between the 'dees'

28 Multiple correct answers

29 a) Acceleration refers to the change in velocity of the particle in an accelerator in order to reach speeds near that of light. **b)** Deflection is the use of a magnetic field to change the direction of a beam, in the case of a synchrotron, to bend it round the circle. **c)** A collision in a particle accelerator is where two or more particles hit each other, usually in a particle collision detector.

30 Multiple correct answers, but they should be based around the observation of the effects of the particles.

Chapter 11: Wave properties

1

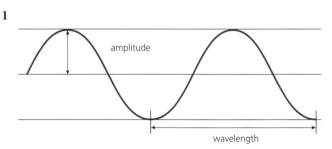

2 $100\,\mathrm{m\,s^{-1}}$

3 10×10^9 or $10\,\mathrm{GHz}$

4 $0.1\,\mathrm{ns}$

5 $7.5 \times 10^{14}\,\mathrm{Hz}$ to $6.67 \times 10^{14}\,\mathrm{Hz}$

6 $600\,\mathrm{nm}$ to $625\,\mathrm{nm}$

7 Multiple answers possible

8 Transverse

9 Reflection

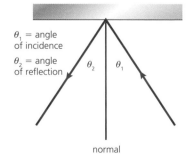

θ_1 = angle of incidence
θ_2 = angle of reflection

normal

10

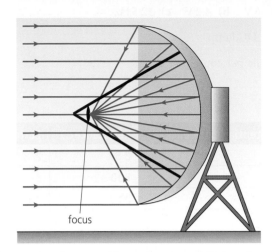

focus

11 A wave is coherent with another wave if it has the same frequency, velocity and wavelength, i.e. a constant phase difference.

12 A constant phase relationship means that the phase difference (the difference between the same point on two separate waves) is constant over time.

13 a) Constructive interference is where two waves combine making a greater amplitude than either original wave.

  b) Two waves add together to give a smaller amplitude.

14 a)

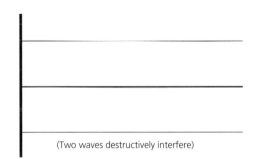

(Two waves destructively interfere)

  b)

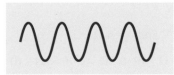

15 a) ii, iii and iv   **b)** (ii) $0.5\,\lambda$, (iii) $0.25\,\lambda$, (iv) $0.125\,\lambda$ **c)** ii

16 0

17 Phase difference is the difference between the same point on two separate waves, the path difference is the difference in how far each of the waves has travelled.

18 Multiple answers possible

19 1.5 m

20 1 m

21 a) Constructive interference, maximum

  b) Destructive interference, minimum

  c) Mixed interference, not a maximum or minimum

  d) Constructive interference, maximum

  e) Destructive interference, minimum

  f) Mixed interference, not a maximum or minimum

22 a) 25 cm   **b)** 0.2 cm   **c)** 10.3 m   **d)** 0.25 μm

23 a) 3 cm   **b)** 6 m   **c)** 4 μm   **d)** 92 mm

24 a) 180 mm   **b)** 10 μm   **c)** 20 m   **d)** 374 nm

25 439 Hz

26 Light of a single frequency

27 A single beam of white light, aimed at a diffraction grating will result in a full visible light spectrum.

28 See Figure 11.43 on page 133. Provided the screen is far away from the sources, the beams will be almost parallel to each other, making the angles shown (approximately) the same.

29 a) 461.7 nm   **b)** 678.7 nm   **c)** 575.5 nm   **d)** 656 nm

30 a) 1.49 μm   **b)** 3.45 μm   **c)** 3.33 μm   **d)** 4.1 μm

31 a) 7°   **b)** 10°   **c)** 1.5°   **d)** 0.015°

32 2 μm

33 Using horizontal and vertical gratings

Chapter 12: Refraction of light

1 See Figure 12.2 on page 141.

2 The wavelength of a wave changes how much it will diffract in a given medium, as a result different colours of light will spread out – we call this dispersion.

3 a) 1.24   **b)** 1.90   **c)** 1.69   **d)** 1.15

4 A = 2.31, B = 14.4, C = 293, D = 52.3, E = 576, F = 3.0×10^8, G = 29.7, H = 543, I = 3.0×10^8, J = 1.8×10^8, K = 1.88, L = 29.7, M = 3×10^8, N = 1.6×10^8

5 a) 17.1°   **b)** 28.0°   **c)** 20.4°   **d)** 65.9°

6 a) 650.1 nm **b)** 422 nm **c)** $1.95 \times 10^8\,\text{m s}^{-1}$
d) 14.1°

7

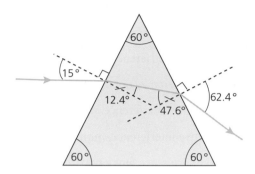

8 a) As the cube is 4 cm long, you can make a right-angled triangle with a side 3 cm long and a side 4 cm long, allowing you to state the angle of refraction as the beam goes into the glass block – 36.9°. The refractive index is 1.5, so you want the angle of incidence to be 64.2°.

b) The diagram should show light entering at an unnamed angle, then going through the glass block at a shallower angle, then leaving the glass block at an angle parallel to the original beam (an exactly upside-down version of Figure 12.4 on page 142).

9 Answer should include reference to dependence of refraction on wavelength or frequency.

10 Answer should include reference to prisms and interference patterns.

Chapter 14: Electrons and energy

1 720 C

2 3.125×10^{20} electrons

3 a) Electrons move in one direction on average.

b) Electrons move back and forth; they do not travel far from their original position.

4

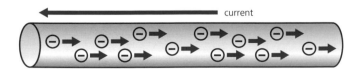

5 a) 156 V

b) The 110 V is V_{rms} whereas the student is reading the peak-to-peak value from the screen.

6 a) 6 V **b)** 4.2 V **c)** 12.5 Hz

7 The DC lamp

8 1 joule per coulomb

9 5 A

10

Potential difference/V	Resistance/Ω	Current/A
5	2.5 k	2×10^{-3}
5	**800**	5×10^{-2}
22	27.5	0.8
230	23	**10**
1	2	0.5
0.5	5000	0.1×10^{-2}
100	80	1.25

11 a) 30 Ω **b)** 66 kΩ **c)** 22.8 kΩ **d)** 1120 Ω

12 a) 0.5 Ω **b)** 2.7 kΩ **c)** 90 Ω **d)** 231 kΩ

13 a) 7.5 Ω **b)** 12.7 MΩ **c)** 52.5 kΩ **d)** 2.8 kΩ

14 a) 100 Ω **b)** 66.7 kΩ **c)** 40.6 MΩ **d)** 800 Ω

15 a) 50 V **b)** 125 V **c)** 37.5 kV **d)** 4.8 V

16 a) 4 mA **b)** 0.18 A **c)** 5 A **d)** 0.3 μA

17 a) 2.5 V **b)** 1.25 V **c)** 4 V **d)** 5.7 V

18 a) 20 V **b)** 11 V **c)** 11.25 V **d)** 7.3 V

19 6 Ω

20 Two 1 kΩ in series with two 1 kΩ in parallel. The voltmeter should be across either of the series resistors.

21 a) 3.1 W **b)** 2 W **c)** 12.5 W **d)** 19.5 W

22 a) All four lamps in parallel. If the lamps are in parallel, each has the maximum p.d. across it – meaning that current through it is as high as possible.

b) 24 W

23 $E = 4.7\,\text{V}$ and $r = 1\,\Omega$

24 a) 0.41 Ω **b)** 0.5 Ω **c)** 0.33 Ω **d)** 2.5 Ω

25 a) 6.1 V **b)** 10.5 V **c)** 3.1 V **d)** 50.3 V

26 8.99 Ω

27 a) Either the capacitance or the resistance should be reduced.

b) The capacitance of the capacitor or the supply voltage must be increased. The resistor should be decreased in size either way in order to maintain the time constant.

28 **a)** 60 pC   **b)** 0.5 mC   **c)** 0.2 C   **d)** 54 mC

29 The voltmeter reading does not go up as expected. The supply in this circuit is AC, so the electric field across the capacitor changes constantly and electrons do not build up on the plates. The supply must be changed to a DC supply.

30 **a)** Simple time-constant curve on a graph. The maximum potential should be 12 V.

b) The circuit should now take longer to charge/discharge.

c) The circuit will now take a longer time than a) and a shorter time than b).

d) The circuit should take the same time as a) but reach double the potential.

e) The graph should no longer be curved, but instead should charge and discharge in a straight line, more quickly than part a).

31 **a)** 14.4 mJ   **b)** 180 pJ   **c)** 5 mJ   **d)** 205 μJ

Chapter 15: Electrons at work

1 See Figure 15.2 on page 201 – conduction and valence bands cross over, electrons are free to move in conductors.

2 Semiconductors, in general, increase conductivity as temperature rises. This is because the thermal energy allows more electrons to jump from the valence band to the conduction band.

3 See Figure 15.20 on page 209. Part b) = no bias; part c) = forward bias; part d) = reverse bias

4 A single solar panel with a voltmeter across it. Angle and irradiance of each colour of light must be the same, and there must be a range of wavelengths/frequencies.

Answers to consolidation questions

Chapter 1: Motion

1 A vector has a size (magnitude) and direction. A scalar has only magnitude.

2 Scalars – Any two of: energy, distance, mass, speed, power

Vectors – Any two of: force, displacement, acceleration, momentum

3 Average speed $= 7\,km\,h^{-1}$

Average velocity $= 5\,km\,h^{-1}$

Direction $= 53.1°$ East of North

4 $2.96\,m\,s^{-2}$

5 $200\,m$

6 a)

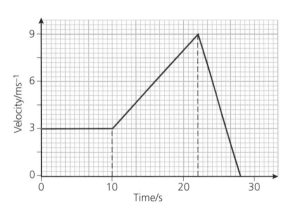

b)

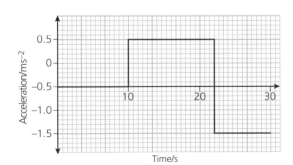

7 a) $6.1\,m\,s^{-1}$ **b)** $0.62\,s$ **c)** $1.27\,m$ **d)** $0.51\,s$

e)

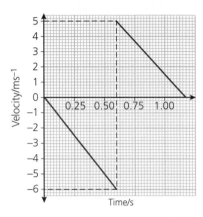

8 a) $1.53\,s$ **b)** $0\,m\,s^{-1}$

c)

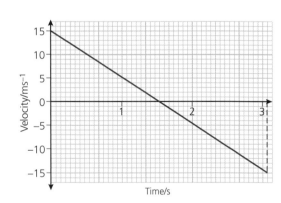

d)

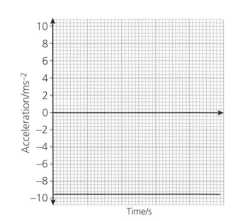

9 Velocity $= 136\,m\,s^{-1}$; at a bearing of $009°$

10 a) $8.6\,m\,s^{-1}$ **b)** $6.5\,m\,s^{-1}$

11 Displacement $= 228\,m$ at a bearing of $071°$

12 a) (i) $0.25\,m$ **(ii)** $2.25\,m$

(iii) $s = ut + \frac{1}{2}at^2$

$s = \frac{1}{2}at^2$

$a = \frac{2s}{t^2} = \frac{2 \times 2.25}{(0.77)^2} = 7.6$

b) The peak would be lower and the time would be greater; there would be a more elongated curve along the time axis.

13 a) (i) 1 m 0.60 s; 2 m 0.85 s

(ii) 3.33 (or 3.36 depending upon equation used and rounding) m s^{-1}; 4.8 m s^{-1}

(iii) 4.1 m s^{-1}. It is the same. The mass undergoes a constant acceleration therefore the increase in velocity is the same for each increment. It is a linear increase.

14 a) (i) 3.9 m s^{-2} (ii) 60 m s^{-2}

Chapter 2: Forces, energy and power

1 a) $Fd = m \times g \times \sin 42° = 12 \times 9.8 \times \sin 42° = 78.7 = 79$ N

b) Force causing acceleration $= m \times a = 12 \times 6.0 = 72$ N
Friction $= 79 - 72 = 7$ N

2 The ball leaves the bench with horizontal kinetic energy and gravitational potential energy. It accelerates downwards as its gravitational potential energy is transformed to kinetic energy (in the downward direction). The ball rebounds and its vertical kinetic energy is now being transformed to potential energy; however, the collision with the floor was not elastic. During the collision, some kinetic energy was converted to heat and sound. The ball now has slightly less kinetic energy and as a result will not reach the same height nor will it have the same horizontal velocity.

3 a) 1.1 m s^{-2} b) 3.7 s

4 a) (i) 2.0 m s^{-2} (ii) 961 m (960 m)   (iii) As the aircraft increases in velocity, the air friction increases.

b) (i) 42 700 N   (ii) The lift is perpendicular to the line of flight.   (iii) 165 000 N

5 a) (i) $6 \times \cos 25° = 5.4$ N   (ii) 0.25 m s^{-2} (iii) 9.9 s

b) Time will be less as the horizontal component of the force will be greater due to a smaller angle. This means a larger acceleration and thus less time.

6 a) 71 m   b) velocity 3.4 m s^{-1}; speed 7.9 m s^{-1}
c) 0 m s^{-1}

Chapter 3: Collisions, momentum and energy

1 a) Horizontal $= 150.4$ m s^{-1}; vertical $= 54.7$ m s^{-1}

b) (i) 1.9 m s^{-1} (ii) 71.6 m s^{-1} (iii) 26 800 N
(iv) Various answers possible

2 a) 0.046 J and 0.061 J   b) The metal spring experiment, as little kinetic energy has been lost   c) Pull back on the elastic band for the same distance, then release.

3 a) (i) 1.7 m s^{-1} (ii) 400 m s^{-1} b) The height would be greater as more momentum is transferred.

4 a) (i) 6.3 m s^{-1} (ii) 4700 N (downward direction)

b) The length of time that the two are in contact increases, so the force decreases.

5 a) 0.41 m s^{-1} b) (i) Space probe's engine   (ii) 154 s
c) The vehicle fires the engine for approximately 2 s and then stops. The craft moves at a steady velocity. The probe fires the engine for approximately 4 s to slow down and stop.

6 a) 90 kg m s^{-1} b) 0.75 s   c) 0.86 m s^{-1}
d) No, because E_k before $= 312.5$ J and E_k after $= 476.1$ J

Chapter 4: Projectiles and satellites

1 a) 1 m $= 0.45$ s; 2 m $= 0.64$ s; 3 m $= 0.78$ s; 4 m $= 0.9$ s

b)

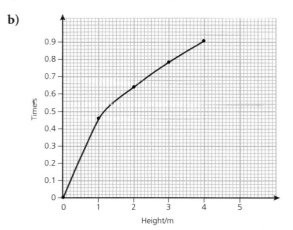

c)

$s = \frac{1}{2}at^2$

$t = \sqrt{\frac{2s}{a}}$

$t \therefore = \sqrt{s}$ This is a curved graph.

2 a) 1.22 s   b) 1.37 s   c) 13.5 m s^{-1}

3 a) 7800 m s^{-1}; 7640 m s^{-1}; 7350 m s^{-1}; 5930 m s^{-1}; 4930 m s^{-1}; 3860 m s^{-1}; 3060 m s^{-1} b) Various answers possible

4 a) 718N **b)** $12\,\mathrm{m\,s^{-1}}$ **c)** 3.3s

5 a) 0.61s **b)** $11.5\,\mathrm{m\,s^{-1}}$

6 a) (i) $2.2 \times 10^6\,\mathrm{N}$ **(ii)** $0.09\,\mathrm{m\,s^{-2}}$

b) There is greater resistance from the water as the rig speeds up.

7 a) Inelastic **b)** 0.21m

8 C

9 a) (i) $v^2 = u^2 + 2as$
$$0 = u^2 + 2as$$
$$u = \sqrt{-2as}$$
$$= \sqrt{(-2) \times (-9.8) \times 1.0}$$
$$= 4.4\,\mathrm{m\,s^{-1}}$$

(ii) $9.0\,\mathrm{m\,s^{-1}}$

b) It was greater. If his height is lower then he is in the air for less time, therefore his speed needs to be greater.

Chapter 5: Special relativity

1 a) $11\,\mathrm{m\,s^{-2}}$ **b)** 10m **c)** 110m

2 $c = 1/\sqrt{(\varepsilon_o \times \mu_o)} = 1/\sqrt{(8.854 \times 10^{-12} \times 1.257 \times 10^{-6})}$
$= 2.998 \times 10^8\,\mathrm{m\,s^{-1}}$

3 a) The speed of light is a constant value in every frame of reference. Galilean invariance holds true for all frames of reference. **b)** Physical laws established in the laboratory hold true universally and are not dependent on where they are measured or how the observer might be in motion whilst making the observation.

c) (i) Even though the separation of the two spacecraft is increasing at a rate greater than c, it is not a real object and so the speed of separation increase can exceed the 'universal speed limit' of c. It is a stationary observer who makes the measurement of both craft from their starting point and uses the two measurements to derive the rate of change of separation. **(ii)** It would not be possible for an observer on one spacecraft to measure the speed of the other spacecraft. Information exchange between the craft can travel only at the speed of light and as the separation is growing in excess of this rate then light from one craft would not be able to 'catch up' with the other craft.

4 For $v = 0.98\,c$, $\gamma = 1/\sqrt{(1 - (v/c)^2)}$, so $\gamma = 5$

a) $t' = \gamma t$, so the stationary observer measures the half-life as $t' = 5 \times 1.56 \times 10^{-6} = 7.8 \times 10^{-6}\,\mathrm{s}$

b) To the stationary observer the journey is 10km at $0.98\,c$ so takes $34 \times 10^{-6}\,\mathrm{s}$. This is just over 4 half-lives (actually 4.36). The number of remaining muons is given by $N = 1 \times 10^6/2^{4.36} \approx 49\,000$

c) $l' = l/\gamma$, so the muon-based observer measures the distance travelled as $l' = 10\,\mathrm{km}/5 = 2\,\mathrm{km}$

Chapter 6: The expanding Universe

1 $f_o = f_s\,(v/(v + v_s)) = 277 \times (340/(340 + 8)) = 271\,\mathrm{Hz}$
on approach

$f_o = 277\,\mathrm{Hz}$ at the point the horn is level with the observer

$f_o = f_s\,(v/(v - v_s)) = 277 \times (340/(340 - 8)) = 284\,\mathrm{Hz}$
after passing observer

2 a) $f_o = f_s\,(v/(v + v_s))$ so $480/440 = 340/(340 - v)$;
$v = 28.3\,\mathrm{m\,s^{-1}}$

b) (i) $\lambda_{rest} = v/f = 340/440 = 0.77\,\mathrm{m}$

ii) $\lambda_{observed} = v/f = 340/480 = 0.71\,\mathrm{m}$

3 $f_o = f_s\,(v/(v - v_s))$ so $7002.5/7000 = 1540/(1540 - v)$;
$v = 0.55\,\mathrm{m\,s^{-1}}$

4 a) Observed wavelength is shorter so the spectrum is blue-shifted, i.e. galaxy is approaching.

b) $z = (\lambda_{observed} - \lambda_{rest})/\lambda_{rest} = (633 - 656)/656 = -0.035$

c) $z = v/c$; $0.035 = v/3 \times 10^8$; $v = 1.05 \times 10^6\,\mathrm{m\,s^{-1}}$

d) $v = H_o d$; $1.05 \times 10^6 = 2.27 \times 10^{-18} \times d$;
$d = 4.63 \times 10^{23}\,\mathrm{m}$

5 $v = H_0 d$. Relative Velocity of Star, Hubble Constant, Distance to the Star. For units see text. Hubble's original data was the redshift values for Cepheid stars calculated in the 1920s.

6 $\mathrm{km\,s^{-1}\,Mpc^{-1}}$ are kilometres per second per Megaparsec! You can convert the value of H_0 to SI units as follows. For example, take the Hubble constant H_0 to be $70\,\mathrm{km\,s^{-1}\,Mpc^{-1}}$ and one light year to be $9.46 \times 10^{15}\,\mathrm{m}$. One Parsec $= 3.26$ light years $= 3.0857 \times 10^{16}\,\mathrm{m}$, therefore 1 Mpc $= 3.0857 \times 10^{22}\,\mathrm{m}$.

So $70 \, \text{km} \, \text{s}^{-1} \, \text{Mpc}^{-1} = 70 \times 10^3 / 3.09 \times 10^{16} \times 10^6$
$= 2.27 \times 10^{-18} \, \text{s}^{-1}$. Given the speeds and distances involved, the non-SI units are more appropriate and give rise to a more manageable number, i.e. 70 rather than $\times 10^{-18}$.

7 See answer 6 for conversion of H_0 to SI units.

a) $500 \, \text{km} \, \text{s}^{-1} \, \text{Mpc}^{-1} = 500 \times 10^3 / 3.09 \times 10^{16} \times 10^6$
$= 1.62 \times 10^{-17} \, \text{s}^{-1}$

$t = 1/H_0 = 1/1.62 \times 10^{-17} = 6.18 \times 10^{18} \, \text{s}$

$72.5 \, \text{km} \, \text{s}^{-1} \, \text{Mpc}^{-1} = 72.5 \times 10^3 / 3.09 \times 10^{16} \times 10^6$
$= 2.34 \times 10^{-18} \, \text{s}^{-1}$

$t = 1/H_0 = 1/2.34 \times 10^{-18} = 4.27 \times 10^{17} \, \text{s}$

b) The SI calculation gives units of seconds; converting to years gives the more familiar values of 1.96 & 13.6 billion years respectively.

c) Hubble's work used the best technology available at the time but could only examine the very closest stars. Subsequent advances in technology have allowed astronomers to look at very, very distant stars and galaxies and have therefore improved the accuracy of the result.

Chapter 7: The Big Bang Theory

1

Type	Temperature/K	Wavelength/m	Peak Colour
O	25000	1.16×10^{-7}	Far ultraviolet
B	17500	1.66×10^{-7}	Far ultraviolet
A	8750	3.31×10^{-7}	Near ultraviolet
F	6750	4.29×10^{-7}	Violet
G	5500	5.27×10^{-7}	Green
K	4250	6.82×10^{-7}	Infrared
M	3500	8.28×10^{-7}	Infrared

2 See the text in Chapter 7

Chapter 8: Electric fields

1 a) $88 \, \text{kV}$

b) The wide spread of paint is because the droplets repel each other as they all have the same charge. The reduction in waste is because the negative charge of the car attracts the positively-charged paint, reducing the number of droplets which fall on the ground.

2 a) i) $2.4 \times 10^{-16} \, \text{J}$

ii) $2.3 \times 10^7 \, \text{m} \, \text{s}^{-1}$

b) P plate 2 and Q plate 2

Chapter 9: The Standard Model

1 a) 91   b) 39   c) 44

2 $1.6 \times 10^{-19} \, \text{C}$; $300\,000\,000 \, \text{m} \, \text{s}^{-1}$; $1.5 \times 10^{11} \, \text{m}$; $0.000\,000\,000\,000\,001 \, \text{m}$

3 See Figure 9.1 on page 109.

4 Baryon: compound particle consisting of three quarks (either all matter quarks or all anti-matter quarks). Using this rule a great many baryons may be postulated. They are grouped as Delta, Lambda, Sigma, Chi and Omega baryons with subtle naming changes based on the constituent quarks. Remember that for each matter baryon there is a corresponding anti-matter one.

5 Meson: compound particle consisting of two quarks (one must be matter, the other an anti-matter quark). Using this rule a great many mesons may be postulated. Their naming is more complicated and some are particularly obscure (e.g. Omega meson). They have their own anti-matter particle but some have a distinct anti-matter partner!

Chapter 10: Nuclear reactions

1 Mass before: $3.88638 \times 10^{-25} \, \text{kg}$

Mass after: $3.81986 \times 10^{-25} \, \text{kg} + 6.64647 \times 10^{-27} \, \text{kg}$
$= 3.88632 \times 10^{-25} \, \text{kg}$

Mass difference: $5.53 \times 10^{-30} \, \text{kg}$

Assuming all mass difference energy is transferred to the alpha particle kinetic energy:

$E = mc^2 = 5.53 \times 10^{-30} \times (3 \times 10^8)^2 = 4.977 \times 10^{-13} \, \text{J}$

2 a) $^{236}_{92}\text{U} \rightarrow ^{90}_{38}\text{Sr} + ^{136}_{54}\text{Xe} + 12^1_0\text{n}$

b) Mass before: $3.91974 \times 10^{-25} \, \text{kg}$

Mass after: $1.45971 \times 10^{-25} \, \text{kg} + 2.25679 \times 10^{-25} \, \text{kg}$
$+ 12 \times (1.67492 \times 10^{-27}) \, \text{kg} = 3.91749 \times 10^{-25} \, \text{kg}$

Mass difference: $2.2496 \times 10^{-28} \, \text{kg}$

Energy release: $E = mc^2 = 2.2496 \times 10^{-28} \times (3 \times 10^8)^2 = 2.025 \times 10^{-11} \, \text{J}$

3 See Figure 10.3 on page 115.

Chapter 11: Wave properties

1 a) (i) The maxima are heard when the pupil is at a point where the path difference is equal to exactly one wavelength. The maxima will also occur when the path difference is 0λ, 2λ, etc. (ii) The minima are where the path difference is equal to exactly one half wavelength. Minima will also occur at $\lambda/2$, $3\lambda/2$, etc. b) The light sources may not be coherent.

2 a) $(2.403 \pm 0.004)\,\text{m}$ b) x (2.2%)
c) $(653 \pm 14)\,\text{nm}$ d) Measure x to the second order maximum.

3 a) 477 nm b) Blue

4 a) Answer should include reference to constructive and destructive interference. b) 24 mm

5 a) 101 lines per mm b) Shorter wavelength

Chapter 12: Refraction of light

1 The critical angle is the angle of incidence at which refraction will occur at exactly 90°. If the angle of incidence is increased total internal reflection will occur.

2 See Figure 12.18 on page 146. Total internal reflection occurs when the angle of incidence is greater than the critical angle.

3 a) 52.5° b) 24.5° c) 32.3°

4 a) 1.65 b) 1.05 c) 1.39

5 a) Critical angle is 34.8°; refraction occurs; TIR does not occur. b) Critical angle is 36.0°; TIR occurs. c) Critical angle is 45.2°; TIR occurs. d) RI is 1.2; critical angle is 56.4°; TIR does not occur.

6 See Figure 12.19 on page 146.

7 In a copper cable, some energy is used heating the copper – this reduces the signal strength. In a fibre optic cable, very little energy is lost as the signal propagates down the line.

8 a)

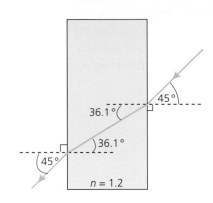

b)

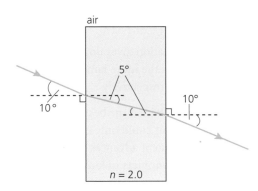

c)

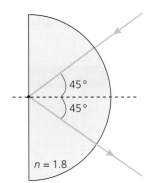

d)

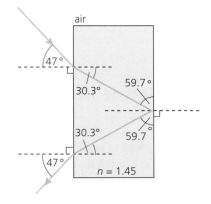

Chapter 13: Spectra

1 a)

Quantity	Name	Unit	Symbol
E_k	Kinetic energy of ejected electron	Joules	J
hf	Incoming photon energy	Joules	J
hf_0	Energy required to eject an electron from its parent atom, also called the work function	Joules	J
f_0	Threshold frequency, i.e. the minimum frequency of a photon which causes electron emission	Hertz	Hz

b) See Figure 13.3 on page 151. **c)** The gradient of the graph of E_k versus f gives the value for Planck's constant, h. **d)** A line parallel to the first line (i.e. same gradient) but shifted to the right so the y-axis intercept (hf_0) is larger.

2 a) The energy of each incoming photon ($E = hf$) is larger than the work function and so can cause the electron to be ejected from its parent atom.

b) $hf = hf_0 + E_k$ so $hf_0 = hf - E_k$

$$= 6.63 \times 10^{-34} \times$$
$$(3 \times 10^8 / 454 \times 10^{-9})$$
$$- 3.36 \times 10^{-19}$$
$$= 1.02 \times 10^{-19} \, \text{J}$$

c) The maximum kinetic energy is unchanged by the power of the light source. As shown by the equation, the kinetic energy is the difference between the incoming photon energy and the work function of the metal.

3 Any three of the following:

◆ There was a measurable minimum potential difference which stopped all electrons reaching the collector and this did not depend on the irradiance of the illuminating light.

◆ Doubling the intensity similarly doubled the number of electrons emitted (i.e. the current measured) but had no effect on the kinetic energy of the emitted electrons.

◆ The maximum kinetic energy of the electrons emitted depended on the frequency of light being used to illuminate the metal, i.e. higher frequency meant higher kinetic energy.

◆ There was a minimum value of light frequency below which no electrons were emitted.

◆ There was no delay in the emission of electrons. Even for very low light levels, electrons were emitted immediately if the frequency was sufficiently high.

4 The inverse square rule for point source irradiance $I = k/d^2$ may be used to give:

$$I_1 d_1^2 = I_2 d_2^2$$
$$6 \times 10^7 \times (6.6 \times 10^8)^2 = I_2 \times (1.5 \times 10^{11})^2$$
$$I_2 = 1.3 \times 10^3 \, \text{W m}^{-2}$$

5 a) See Figures 13.5 and 13.6 on pages 152 and 153.

6 486.2 nm; 6.17×10^{14} Hz; 6.91×10^{14} Hz; 4.58×10^{-19} J; 410.4 nm; 4.84×10^{-19} J

7 a) This is $n = 1$ to $n = \infty$ transition, so ionisation energy $= 2.179 \times 10^{-18}$ J

b) $E = hf = E_4 - E_2$

$$hf = -1.362 \times 10^{-19} - (-5.447 \times 10^{-19})$$
$$f = 4.085 \times 10^{-19} / 6.63 \times 10^{-34}$$
$$f = 6.16 \times 10^{14} \, \text{Hz}$$

so $\lambda = 3 \times 10^8 / 6.16 \times 10^{14} = 486$ nm which is a blue-green colour.

c) 434.1 nm gives $f = v/\lambda = 6.91 \times 10^{14}$ Hz

Therefore $E = hf = 6.63 \times 10^{-34} \times 6.91 \times 10^{14} = 4.58 \times 10^{-19}$ J. This is the $n = 5$ to $n = 2$ transition.

d) The lowest energy transition to the ground state is $n = 2$ to $n = 1$, which corresponds to an energy

$$E = hf = E_2 - E_1$$
$$hf = -5.447 \times 10^{-19} - (-2.179 \times 10^{-18})$$
$$f = 1.634 \times 10^{-18} / 6.63 \times 10^{-34}$$
$$f = 2.47 \times 10^{15} \, \text{Hz}$$

so $\lambda = 3 \times 10^8 / 2.47 \times 10^{15} = 121$ nm, which is a deep ultraviolet and so is invisible to humans.

Chapter 14: Electrons and energy

1 $21 \, \Omega$

2 $8 \, \Omega$

3 $12 \, \text{V}$

4 a) **(i)** $5.34 \, \text{k}\Omega$ **(ii)** As the total parallel resistance of Z and Y is less than the resistance of Y alone, V_0 gets smaller.

(iii) 5.77 V

b) $R_A / R_B = R_C / R_D$

c)

Resistance of A/Ω	Resistance of B/Ω	Voltmeter reading/mV
60	60	**0**
61	60	-21
61	61	**0**
61	62	$+41$
61	59	-83

5 a) 14.8 V

b) The diode stops the discharging current from flowing.

6 a) (i) 400 kHz   (ii) 12V   (iii) 8.5 V

b) (i) Charging time is doubled   (ii) No effect on energy stored; however, the capacitor will reach this energy in half the time.   (iii) Maximum charge is doubled but charging time is the same (halved because of the resistance, but doubled because of the capacitance).

Chapter 15: Electrons at work

1 a) (i) 4.1 mA   (ii) 13.4 mJ   (iii) Decrease the supply   **b)** 1.20×10^{16} photons

2 a) 4.5 J of energy is provided per coulomb of charge

b) (i) 13.5 Ω

(ii) 4.28 V

c) The overall resistance reduces, the current increases, the lost volts increase so the voltmeter reading reduces

3 a) The resistivity reduces due to doping   **b)** (i) When electrons fall from the conduction to valence band (to recombine with holes) their energy is reduced. This energy is released as a photon.   (ii) 3.7×10^{-6} m

4 a) 5.6 Ω **b)** 0.54 A   **c)** 1.4 J   **d)** The power dissipated will increase as the total resistance decreases, meaning the current in the element increases.